530.11

Kahan, Gerald

$E = mc^2$

No. 1580
$16.95
-5

$$E = mc^2$$

# PICTURE BOOK
# OF RELATIVITY

## By Gerald Kahan
## Illustrated by Charles Prodey

**TAB** **TAB BOOKS Inc.**
BLUE RIDGE SUMMIT, PA. 17214

*The whole of science is nothing more than a refinement of everyday thinking*

A. Einstein

FIRST EDITION

FIRST PRINTING

Library of Congress Cataloging in Publication Data

Kahan, Gerald, 1942-
E = mc$^2$ : picture book of relativity.

Includes index
1. Relativity (Physics)—Popular works. I. Prodey,
Charles, 1945-   . II. Title.
QC173.57.K33   1983     530.1'1    83-10573
ISBN 0-8306-0280-1
ISBN 0-8306-0180-5 (pbk.)

Cover photograph by Charles Prodey.

# Contents

With
Love and Affection
to

Barbara
Jeffrey
Cindy

and

Virginia

# Preface

Work on this book started over 25 years ago. At that time, a trip to the Hayden Planetarium in New York City sparked my lasting interest in astronomy. Albert Einstein was still alive then, and it seemed like everything he said or did made headlines. I was intrigued. I began with a few books on relativity taken from the juvenile section of the library, but it was hopeless! I was simply too young to understand it. Despite my disappointment, I resolved to come up with an explanation of relativity that could be explained to young people. Every few years, I would take books on relativity out of the library, pore over them, and haul them back to the library with a feeling of frustration. The concepts seemed incomprehensible. Several years ago, while having dinner with my family, I had one of those rare moments of insight when it all fell into place.

I put my ideas together in the form of a lecture on relativity. For several years, I presented this lecture as part of a Student Science Seminar program sponsored by the Maryland Academy of Sciences. I also presented the talk to other interested groups including the Baltimore Astronomical Society and the Baltimore Chapter of Mensa. With each lecture, I made modifications to the presentation in response to questions and comments. What has evolved from these experiences is presented in the pages of this book.

# Introduction

In developing a simple explanation of relativity, I found that I had to avoid a concept which lies at the very heart of Einstein's ideas. This is the concept of simultaneity, a concept which Einstein used to derive his equations. Most popular writers on relativity use simultaneity in their explanations, and, in doing so, present arguments closer to those actually used by Einstein. Although I have assumed that the reader has an understanding of some of the basic concepts of science, I have sought to avoid simultaneity because I find it much too difficult for the non-scientist. The models and explanations which I have used are not rigorous. They cannot be used to derive the equations of relativity, however, they do serve to convey the logic, the order, and the harmony that tie these strange ideas together. Einstein was a man who felt deeply, a man whose scientific creations were the product of an insatiable curiosity and an incredibly accurate intuition. If I have succeeded in conveying a sense of wonder and mystery, I have captured the genius that was Einstein.

As you can plainly see from the title, pictures and illustrations are an integral part of this book. Virtually every concept has been illustrated with one or more drawings to facilitate comprehension and enhance enjoyment. Experience has taught me that regardless of how well Einstein's ideas may be expressed, most people still require additional time to think about them in order to fully comprehend the many subtle details. The pictures are therefore intended to allow you to reflect for a moment and reinforce the explanation with a visual review—even at the risk of what may at times seem like unnecessary repetition.

It is impossible to thank the many people, who, over so many years, have offered helpful comments and suggestions. My special thanks to Ronald L. Barnes, John Adam Moreau, and Marilyn Koeppel for their advice, encouragement, and counsel, and to John P. Thompson for his unfailing support. Also, a special thank you to Barbara J. Sonberg, a physicist who graciously reviewed the manuscript and provided technical advice which influenced both the form and content of the book. I am indebted to Deborah E. Kreider for typing the manuscript, and to Elizabeth Shields and Elizabeth Hanson for providing additional clerical support. I am most grateful to Charles J. Prodey for the photography and the many beautiful illustrations which I believe make this book unique and fun to read.

During the early stages of writing, Mr. Prodey suggested to me that a puppet of Albert Einstein might add an unusual and engaging touch to our efforts. As a result, we were most fortunate in acquiring the enthusiastic support of Mr. Roy Insley, a man whose long career in the theater arts has included the creation of numerous puppets and miniature theater sets. It is Mr. Insley's Albert Einstein which graces the cover and pages of this book.

Finally, and most sincerely, a very special and warm thank you to my wife Barbara, who, over a period of 20 years, has patiently listened to countless lectures on the Theory of Relativity.

# Starlight

**E**=mc² IS A FORMULA THAT HAS ACHIEVED A STA-
tus in the public mind comparable to the first
four notes of Beethoven's Fifth Symphony and the
mysterious smile on Leonardo DaVinci's Mona Lisa
(Fig. 1-1). It is a formula that oftentimes conjures up
an image of a large and ominous mushroom-shaped
cloud because it states that energy (E) equals mass
(m) times the speed of light (c) squared. The letter $c$
is used to denote the speed of light because it stands
for the word "constant."

Just how "constant" the speed of light really is
will be explained in the next few chapters. However, for the moment, it is sufficient to realize that
the speed of light is a very large number (186,000
miles per second) and when this number is squared,
or multiplied by itself, it results in a still larger
number. When this very large number is multiplied
by any number for mass, even a small one, it results
in a very large value for energy. This explains how
with only 110 pounds of uranium, the approximate
amount of uranium used in the atomic bomb dropped
on Hiroshima, there was sufficient energy available
to destroy an entire city.

The formula was first written in 1905—early
enough in the twentieth century so that it would
seem unlikely that anyone, least of all Albert Einstein, should have been thinking of anything even
remotely connected with an atomic bomb. What
then was he thinking about? Like other physicists at
the time he was thinking about many problems, but
the specific problem to which he addressed himself
when he wrote this formula is one which we will
discuss later in this story.

For now, in order to give this formula and the
events leading up to its creation historical perspective, we will address ourselves to a much larger
question which concerned physicists at the time—a
question concerning starlight. Specifically, "How
does starlight manage to travel through the vacuum
of empty space?" How we manage to weave a con-
necting thread between something as tenuous as
starlight and as real as the atomic bomb is a story
that begins as early as the second century A.D.

Although very little is known about his life,
Claudius Ptolemaeus, an Alexandrian Greek now
more widely known as Ptolemy, is believed to have

1

Fig. 1-1.

lived in the second century A.D. (Fig. 1-2). His influence in astronomy, geography, and mathematics was strongly felt for over thirteen centuries. Paramount among his many ideas was the concept of an Earth-centered, or geocentric, universe. Simply stated, Ptolemy maintained that the Earth stood still, while all heavenly bodies, including the planets, revolved around the Earth. Ptolemy enshrined this idea in his 13-volume *He mathematike syntaxis* (The Mathematical Collection), which is today known as the *Almagest*.

Ptolemy advanced several seemingly reasonable arguments to support the idea of a motionless Earth positioned at the center of the universe. He argued that since all bodies fall to the center of the universe (as Aristotle had asserted) and all falling objects are seen to drop towards the center of the Earth, the Earth must be located at the center of the universe. Furthermore, if the Earth rotated, an object thrown vertically upwards would not fall back to the same place as it was in fact seen to do. Consequently, the Earth must be standing perfectly still and not rotating every 24 hours as some had suggested. These ideas, popularly known as the Ptolemaic system, became unassailable dogma during the Middle Ages.

In keeping with the tradition of the Greeks, Ptolemy affirmed the notions of symmetry and harmony by asserting that the planets travel through the heavens along circular paths. In order to explain what everyone had already observed, namely, that as the planets moved across the evening sky they shifted sometimes to the left and sometimes to the right, it was necessary to invent a complicated system of epicycles and deferents. Ptolemy imagined that a planet moved in a circle called an epicycle, while the center of the epicycle traveled around the Earth in an orbit called a deferent (Fig. 1-3). During

2

Fig. 1-2. Ptolemy. Despite his far reaching influence, no one ever wrote a biography or even a popular text describing his work (courtesy of Brown Brothers).

In 1543, a Polish monk named Nicolaus Copernicus struck the first blow at what had become 13 centuries of smug complacency (Fig. 1-4). In that year, he published *De revolutionibus orbium coelestium* (On the Revolutions of the Celestial Spheres), in which he advanced the idea that all of the planets, including the Earth, traveled around a stationary Sun. He also maintained that the Earth revolved around its own axis and followed a circular path in its journey around the Sun. By establishing the idea of a solar system, Copernicus succeeded in eliminating what had become an increasingly awkward and cumbersome system to live with, the system of epicycles.

Nicolaus Copernicus was born in eastern Poland in the year 1473. He attended several universities, including the University of Cracow and the University of Padua, and ultimately earned the degree of doctor of canon law. However, his studies were so wide ranging that he is thought to have mastered all that was then known in mathematics, astronomy, medicine, and theology. By 1514 his fame in the fields of astronomy had increased to the point where he was invited to attend the Lateran Council, a meeting of church authorities considering, among other things, calendar reform.

the course of the centuries that followed, this system became more and more difficult to accept as observations became more and more accurate.

Finally, in Ptolemy's conception of the universe, the planets were closer to the Earth than the "fixed" stars. He believed that the stars were attached to a large crystalline sphere. Outside this sphere were other spheres ending in the *primum mobile* (prime mover), a sphere which provided the motive power for the other spheres.

The Ptolemaic system brought security, comfort, and support to a world dominated by religion and Biblical Scripture. Man was the center of all creation and the universe merely a pale reflection of his existence. As the ancient civilizations crumbled, even the desire to question disappeared. During the Middle Ages, the Bible, and occasionally the teachings of the ancients, provided all of the necessary answers. Man, at least for the time being, was content.

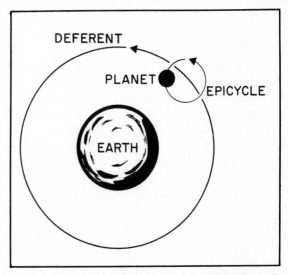

Fig. 1-3. The system of epicycles and deferents. A complicated system invented to explain why the planets sometimes appeared to drift backwards relative to the distant stars.

his ideas were true, Copernicus wisely withheld publication of his views for fear of antagonizing church authorities. A copy of *De revolutionibus orbium coelestium* is said to have been handed to him on the last day of his long life, May 24, 1543.

Ninety years later, in 1633, the seriousness of the Copernican doctrine was underscored by the trial of Galileo Galilei (Fig. 1-5). As the first astronomer to use a telescope, Galileo was able to verify the teachings of Copernicus and prove that the Earth travels around the Sun, and is not the center of the universe. In order to avoid the wrath of the church, Galileo wrote *Dialogo sopra i due massimi sistemi del mondo, tolemaico e copernicano* (Dialogue Concerning the Two Chief World Systems—Ptolemaic and Copernican), a noncommittal discourse between three friends. One friend argues the theories of Aristotle, another sides with Copernicus, and the third defends the teachings of the Church. At first, the book was received with wide acclaim. However, within several months, the ecclesiastical authorities in Rome announced that

Fig. 1-4. Nicolaus Copernicus. A brilliant scholar who managed to evade the wrath of the Church by withholding publication of his views until his death. His great work, "On the Revolutions of the Celestial Spheres" was nevertheless censored by the Church from 1616 to 1835 (courtesy of Brown Brothers).

As his studies in astronomy progressed, Copernicus became increasingly dissatisfied with the Ptolemaic system. The accuracy of astronomical observations had reached the stage where it was becoming extremely difficult to accurately calculate the future position of a heavenly body. Copernicus finally concluded that there must be some simpler, alternative system. His research revealed that even the Greeks had entertained the notion of a Sun-centered or heliocentric universe. When Copernicus incorporated this assumption into his calculations, the result was an aesthetically appealing, although not much simpler, system.

In part, this lack of simplicity is explained by his retention of the idea that the planets moved in uniform, circular motion. Although convinced that

Fig. 1-5. Galileo Galilei. Being the first astronomer to use a telescope, he discovered among other things the craters on the moon and the fact that the Milky Way is made up of millions of stars (courtesy of Brown Brothers).

4

Fig. 1-6. Isaac Newton. His scientific work was so comprehensive and complete that he almost became another Aristotle. In England, his influence was so pervasive that no one did important original work in the subjects he touched for at least a century following his death (courtesy of Brown Brothers).

the book did, indeed, support the Copernican viewpoint, and that the author had merely feigned a noncommittal posture. Accordingly, Galileo was brought before the dreaded Inquisition, forced to renounce his findings, and vow never again to teach or discuss Copernicanism in any way. He was then placed under house arrest, a situation which he endured for the remaining eight years of his life.*

The ideas that Copernicus himself advanced with so much hesitation, led to two major changes in man's conception of the universe. First, since the stars continued to appear "fixed" despite the fact that the Earth was now known to be moving, Copernicus was forced to expand the size of the universe and rightly claim that the stars were too far away for any slight change in position to be detected. Second, if the Earth is moving, it was no longer possible to maintain Aristotle's idea that all objects fall to the center of the universe.

The way was now open for Isaac Newton, who

<hr />

*In November 1980, Pope John Paul II ordered a review of Galileo's trial. The outcome of the review has not yet been announced.

would ultimately write the laws governing falling bodies, and follow it with the concept of universal gravitation (Fig. 1-6). It is no wonder that the ideas advanced by Nicolaus Copernicus, ideas which challenged ancient authority, should today be known as "the Copernican Revolution."

Early in the seventeenth century, at about the same time that Galileo was announcing his epoch-making astronomical discoveries, astronomer and astrologer Johannes Kepler (Fig. 1-7) determined that the planets travel around the Sun, not in circles, but in ellipses (Fig. 1-8). The beginning of a truly accurate understanding of planetary motion was finally at hand. Before long, even the Sun lost its place of eminence in the universe as astronomers came to realize that it was merely another one of a myriad of stars that populate the evening sky.

By the nineteenth century, not only was the Sun regarded as being merely another star, but it was also regarded as a star which moved relative to

Fig. 1-7. Johannes Kepler. A brilliant mathematician who discovered the laws of planetary motion despite a stronger interest in astrology than astronomy (courtesy of Brown Brothers).

5

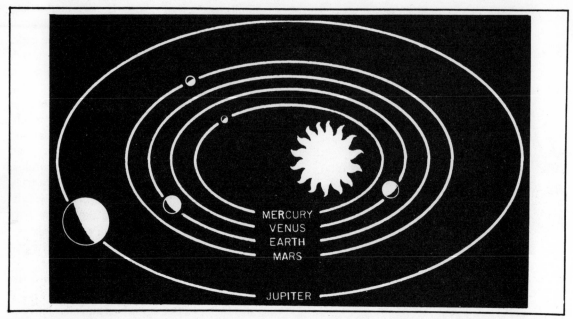

Fig. 1-8. Kepler's first law of planetary motion states: The planets describe elliptical orbits, of which the Sun occupies one focus.

other stars. In fact, it was recognized that every heavenly body—be it a star, a planet, a comet, or whatever—moved in one direction or another about the universe. Absolutely nothing in the universe stood still. Everything moved relative to everything else, a realization that posed yet another problem (Fig. 1-9).

Experimental work by Thomas Young in 1800

Fig. 1-9.

6

Fig. 1-10. Michael Faraday. Although he had very little formal education, he was possibly the world's greatest experimental genius. While working with electricity he developed the "field concept" which later became the basis for Einstein's Theory of Relativity (courtesy of Brown Brothers).

Fig. 1-11. James Clerk Maxwell. Although not well-known to the public, he is familiar to all physicists. He wrote the equations which describe the electric field discovered by Faraday, and thereby established field theory as a very important branch of physics. In fact, field theory became so important that it preoccupied the mind of Albert Einstein for the last 40 years of his life (courtesy of Brown Brothers).

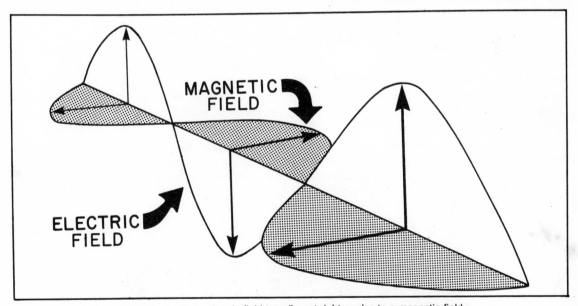

Fig. 1-12. An electromagnetic wave is an electric field traveling at right angles to a magnetic field.

Fig. 1-13. In the nineteenth century, scientists were convinced that a wave required a medium in which to travel.

and Augustin Fresnel in 1814 provided convincing evidence that light consisted of waves; waves not unlike those formed when you drop a ball into a calm lake. Some years later, the brilliant experimental physicist, Michael Faraday, described the effect of an electric field on a magnetic field, and vice versa, by means of lines of force (Fig. 1-10).

In 1862, James Clerk Maxwell, who some would rank with Isaac Newton, extended these findings by expounding the electromagnetic theory of light (Fig. 1-11). According to this theory, light consists of an electric field traveling at right angles to a magnetic field (Fig. 1-12). But if this was true, how then could light travel through the vacuum of empty space? If a ball is thrown into a calm lake, and a bottle is floating about nearby, waves will travel away from the point of impact, past the bottle, causing the bottle to bob up and down. The bottle however will remain in the same relative position in the lake (Fig. 1-13). Even the water near the bottle will move up and down, but the water itself will not move away from the point of impact. The wave will move, but not the water. In order to see a wave, we must watch it as it moves through some medium like water because otherwise there would be nothing to "wave." Nineteenth-century scientists, realizing the need for a medium in order to see a wave, also realized that there was nothing but a vacuum in outer space. Since starlight had to move through some medium, they accepted a suggestion from Maxwell to resurect an old Greek concept called the "ether" and set out to find it.

**Chapter 2**

# Lights and Mirrors

THE CONCEPT OF THE ETHER GREW OUT OF the need to explain how starlight traveled through empty space. The ether was thought to fill all space and penetrate all matter. How else could light pass through empty space as well as a solid pane of glass? The ether was also thought to stand perfectly still while all of the heavenly bodies moved through it. In essence, nineteenth-century scientists saw the universe the way a fish sees the ocean. Wherever a fish turns, other creatures are moving through what appears to be stationary water. In like manner, wherever scientists turned, heavenly bodies were moving through what appeared to be stationary ether (Fig. 2-1). Planets, comets, stars, meteors—all were traveling helter-skelter through an ether thought to be standing perfectly still.

Scientists further reasoned that if man could somehow determine how fast the Earth moved through this stationary ether, man would learn how fast the Earth was moving relative to all creation. By determining this so-called "absolute motion" of

the Earth, man could affirm the fact that the ether really existed after all.

The search for the ether was undertaken in 1879 by a young naval officer named Albert A. Michelson (Fig. 2-2). His work continued on and off for a period of eight years, and culminated in an experiment which he performed in collaboration with a leading chemist of the day, Edward W. Morley (Fig. 2-3). Their collective effort, now known as the Michelson-Morley experiment, was performed in 1887 at the Case School of Applied Science (now Case-Western Reserve University) in Cleveland, Ohio.

To understand the Michelson-Morley experiment, it will be necessary to draw an analogy. Imagine that the wind is blowing on a nice sunny day while you are out watching a friend fly a radio-controlled model airplane. On a day when the wind is not blowing at all, the plane can fly 10 miles per hour. But today the wind is blowing from east to west, and naturally, it is affecting the speed of the airplane. When the plane flies east into the wind, it

Fig. 2-1. Nineteenth century scientists saw heavenly bodies moving through the ether the way fish see themselves moving through water.

Fig. 2-2. Albert Abraham Michelson. Born in Prussia, he grew up in Nevada and California, graduated from the United States Naval Academy, and became the first American to win a Nobel Prize in science (courtesy of Brown Brothers).

travels slower than 10 miles per hour. When it flies west with the wind, it naturally travels faster than 10 miles per hour.

Suppose, for example, that the wind speed was 6 miles per hour. Flying east into the wind, the plane would make headway at 4 miles per hour. Flying west, with the wind, the plane would travel at 16 miles per hour. If you were to time how long it takes for the plane to travel 2 miles east, turn around, and travel two miles west, you would find that it takes ⅝ of an hour, or 37½ minutes (Fig. 2-4).

Suppose now that the air continued to blow from east to west, while the plane flew two miles north, turned around, and flew 2 miles south. You would now find that the trip takes exactly a half hour (Fig. 2-5). On both legs of the journey, the plane would make headway at 8 miles per hour; the fastest possible speed that the plane can move forward, while resisting the force of a wind, which is trying to move the plane west.

What is important here is the fact that it takes longer for the plane to travel against the air current and back (⅝ hour) than it does to travel across the air current and back (½ hour). Although Michelson

Fig. 2-3. Edward Williams Morley. Although he aspired to be a minister, he became a chemist instead and ultimately won fame as a physicist (courtesy of Dr. R.S. Shankland, Professor of Physics, Case Western Reserve University).

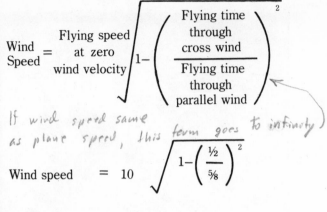

$$\begin{array}{l} \text{Wind} \\ \text{Speed} \end{array} = \begin{array}{c} \text{Flying speed} \\ \text{at zero} \\ \text{wind velocity} \end{array} \sqrt{1 - \left( \dfrac{\begin{array}{c}\text{Flying time}\\ \text{through}\\ \text{cross wind}\end{array}}{\begin{array}{c}\text{Flying time}\\ \text{through}\\ \text{parallel wind}\end{array}} \right)^2}$$

*If wind speed same as plane speed, this term goes to infinity*

$$\text{Wind speed} = 10 \sqrt{1 - \left( \frac{1/2}{5/8} \right)^2}$$

$$\text{Wind speed} = 6 \text{ miles per hour}$$

and Morley had never heard of radio-controlled model airplanes, they used this kind of reasoning to set up their experiment. Before we can describe what they actually did, it will be necessary to develop our analogy a bit further.

With the help of some elementary algebra, it is possible to derive a simple formula to calculate the wind speed using not much more than our airplane and a stop watch. To use the formula, we have only to find the time it takes for the plane to make both round trip journeys and accept the fact that the plane can fly at 10 miles per hour when the wind velocity is zero. We can then substitute these numbers into the formula:

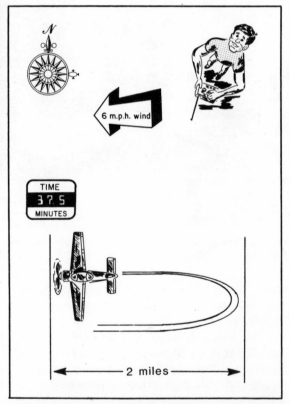

Fig. 2-4. If the wind is blowing at six miles per hour from the east to west, a radio-controlled model airplane (which can fly at 10 m.p.h. when the wind is not blowing) will take 37.5 minutes to travel two miles east into the wind and two miles west with the wind.

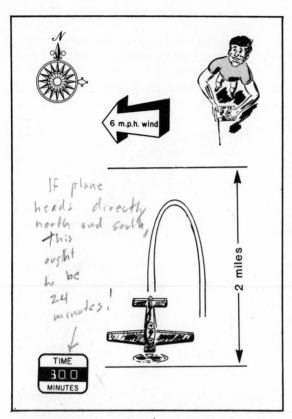

Fig. 2-5. If the wind is blowing at six miles per hour from east to west, a radio-controlled model airplane (which can fly at 10 m.p.h. when the wind is not blowing) will take 30.0 minutes to travel two miles north across the wind and two miles south back across the wind.

Another point to consider is the relative motion of the Earth and the ether. As we have already said, nineteenth-century scientists saw the Earth moving through a stationary ether, just the way we might see ourselves driving along a highway through air that happens to have zero wind velocity. If we are driving at 55 miles per hour, the air will pass us at 55 miles per hour, but if we happen to be standing on the side of the road while someone else drives our car, we will see our car pass through the air at 55 miles per hour. Regardless of whether the air passes the car, or vice versa, the relative speed is 55 miles per hour.

Of course, the same holds true for the Earth and the ether. When we stand on the Earth, we may imagine the ether to be moving past the Earth the way the wind blows past our car (Fig. 2-6). Indeed, scientists spoke of the so-called "ether wind." But if we could stand out in space the way we stand on the side of the road, we could "see" the Earth passing the ether the way our car passes the air (Fig. 2-7). And again, regardless of whether the Earth passes the ether, or vice versa, the relative velocity is the same. This is a fairly obvious, but important point to keep in mind as we develop our analogy still further.

Although Isaac Newton thought that light consisted of particles, which he called "corpuscles," nineteenth-century scientists, as we have already said, showed that light consisted of waves. By the early part of the twentieth century, sometime after the Michelson-Morley experiment was performed, experimental evidence revealed, once again, that light may consist of particles. Einstein called these particles "quanta," and later they became known as "photons." To this day, scientists are unable to say with certainty whether or not light consists of particles or waves. In some experiments it behaves like particles, in others like waves. Scientists therefore speak of the "dual nature" of light. What is important here is not that light may consist of waves or particles, but that we may refer to a particle of light as a "photon" of light. Furthermore, we may think of a photon of light as a little bundle, or ball of energy.

We are finally ready to complete our analogy between the facts concerning the radio-controlled model airplane and the Michelson-Morley experiment. We begin by placing the air with the ether and by replacing the plane with a "slow" photon of light (Figs. 2-8, 2-9, 2-10, and 2-11). Since our plane can travel at 10 miles per hour when the wind is not blowing, our photon can also travel at 10 miles per hour in the stationary ether. Just as the air blows across our face on a windy day, so may we imagine the "ether wind" blowing across our face as the Earth moves through the stationary ether. We said that our plane flies slower than 10 miles per hour when heading east into the wind, and so it should come as no surprise that our photon travels slower than 10 miles per hour when it travels east into the ether wind. We also said that our plane travels

Fig. 2-6. From here on Earth, the ether appears to pass us much like the wind.

faster than 10 miles per hour when flying west with the wind, and so does our photon when it travels west with the ether wind.

If the ether wind is blowing from east to west at six miles per hour, we can expect our photon to travel east at four miles per hour and west at 16 miles per hour. If you were to time how long it takes for the photon to travel two miles east, turn around, and travel two miles west, you would find that it takes ⅝ of an hour or 37½ minutes. If the ether continues to blow from east to west while our photon travels two miles north across the ether current, turns around, and travels two miles south back across the ether current, the entire round trip would take exactly a half hour. Like the airplane, the photon would make headway at eight miles per hour traveling north, and eight miles per hour traveling south, while, at the same time, resisting the force of an ether wind trying to blow it in a westerly direction. Armed with nothing but a stopwatch, we can now determine how fast the Earth is moving through the stationary ether.

We have only to set up an experiment in which we allow a photon to travel two miles east against the ether wind, turn around, and travel two miles west with the ether wind. We can then record the time of the round trip, which we now know will be ⅝ of an hour. We then allow a photon to travel two

miles north across the ether wind, turn around, and travel two miles south back across the ether wind. Again, we record the time of the round trip, which we of course know to be exactly a half hour. With these two stopwatch readings and our knowledge of the fact that the photon travels at 10 miles per hour in a stationary ether, we can substitute into our formula to obtain the speed of the ether wind blowing across the surface of the Earth.

$$\begin{pmatrix} \text{Ether} \\ \text{Wind} \\ \text{Speed} \end{pmatrix} = \begin{pmatrix} \text{Flying} \\ \text{speed at} \\ \text{zero wind} \\ \text{velocity} \end{pmatrix} \sqrt{1 - \left( \dfrac{\substack{\text{Flying time} \\ \text{through} \\ \text{cross wind}}}{\substack{\text{Flying time} \\ \text{through} \\ \text{parallel wind}}} \right)^2}$$

$$\begin{pmatrix} \text{Ether} \\ \text{Wind Speed} \end{pmatrix} = 10 \sqrt{1 - \left( \dfrac{\frac{1}{2}}{\frac{5}{8}} \right)^2}$$

$$\begin{pmatrix} \text{Ether} \\ \text{Wind Speed} \end{pmatrix} = 6 \text{ miles per hour}$$

Since we now know how fast the ether wind blows across the surface of the Earth, and, as previously discussed, we know that the speed of the ether wind is the same as the speed of the Earth

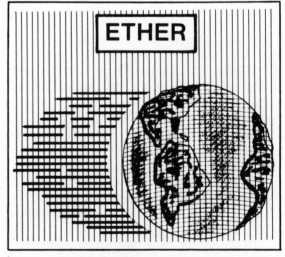

Fig. 2-7. From out in space, the Earth appears to move through the stationary ether.

14

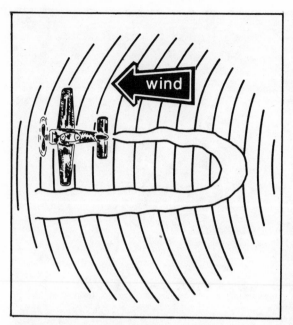

Fig. 2-8. In developing our analogy, we replace the radio-controlled model airplane with a photon of light, and the wind with the ether. Here the airplane is flying east and then west.

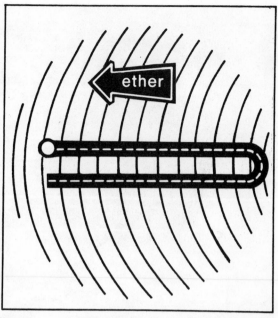

Fig. 2-9. The photon has replaced the airplane and the ether has replaced the wind for the case where the airplane is flying east and then west.

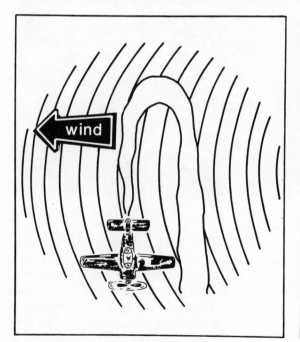

Fig. 2-10. The airplane is flying north and south.

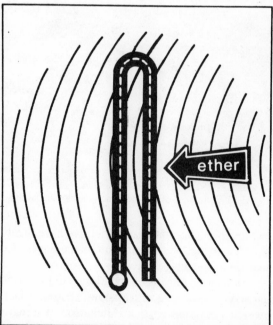

Fig. 2-11. The photon has replaced the airplane and the ether has replaced the wind for the case where the airplane is flying north and then south.

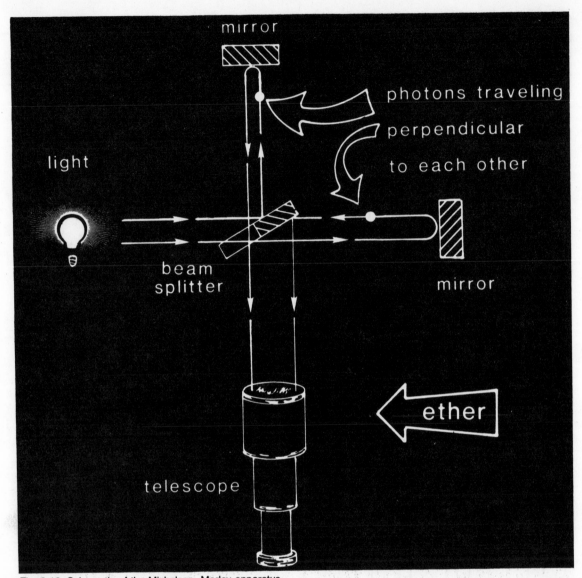

Fig. 2-12. Schematic of the Michelson- Morley apparatus.

moving through a stationary ether, we have finally determined the "absolute motion" of the Earth. Our analogy is complete.

Of course, Michelson and Morley did not have the luxury of working with a "slow" photon. In fact, no one knew better than Michelson that light traveled at approximately 186,000 miles per second. In 1873, when he was only 21 years old, Michelson had performed an experiment revealing

the most accurate measurement of the speed of light made until that time: 186,320 miles per second. The best modern figure is 186,282 miles per second. But, whatever the speed of a photon, the concept embodied in our story of the radio-controlled model airplane can still be carried into the laboratory and used in a real experiment.

What Michelson and Morley actually did was set up an experiment, that is schematically illus-

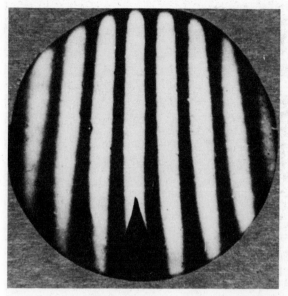

Fig. 2-13. The expected fringe pattern which Michelson and Morley hoped to observe (courtesy of Dr. R.S. Shankland, Professor of Physics, Case Western Reserve University).

trated in Fig. 2-12. Light from a source hit a beam splitter sending two light beams off in directions perpendicular to each other. Each beam traveled to a mirror where it was reflected back, past the beam splitter, to a detector which was actually a small telescope. Michelson and Morley reasoned that once two perpendicular light beams were created at the beam splitter, then, like the travels of the "slow" photon, each would complete the round trip back to the beam splitter in a slightly different amount of time. The beam traveling against the ether current and back would take a tiny fraction of a second longer than the beam traveling across the current and back. The difference in time required for the two beams to make their respective journeys could be computed from measurements of a fringe pattern seen through the telescope.

The fringe pattern, shown in Fig. 2-13, is caused by the reinforcement and interference of light waves. If two light waves are in step, crests and troughs will meet and reinforce each other

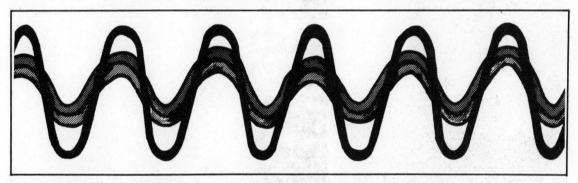

Fig. 2-14. Two light waves which are in step reinforce each other producing more intense light.

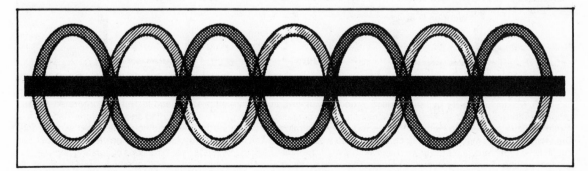

Fig. 2-15. Two light waves which are out of step cancel each other producing no light.

17

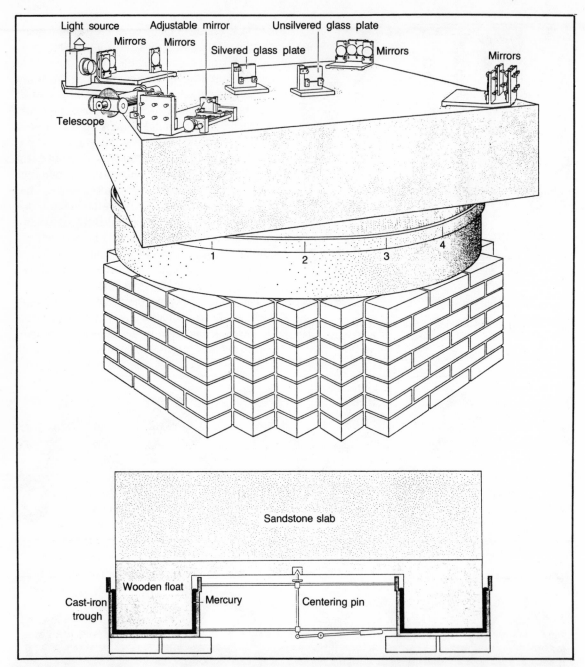

Fig. 2-16. Michelson-Morley apparatus used in decisive ether drift experiment in Cleveland in 1887 incorporated several important improvements over the earlier Michelson interferometers. Optical parts were mounted on a sandstone slab five feet square, which was floated in mercury, thereby reducing the strains and vibrations that had so affected the earlier experiments. The stone itself was mounted on a doughnut-shaped wooden float, which in turn was placed in a similarly shaped cast-iron trough filled with mercury (see cutaway view at bottom). Observations could be made in all directions by rotating apparatus in horizontal plane. (From "The Michelson-Morley Experiment" by Dr. R.S. Shankland. Copyright (November, 1964) by Scientific American, Inc. All rights reserved.)

18

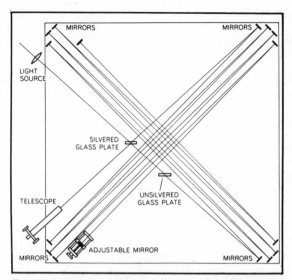

Fig. 2-17. Vertical overhead view of the Michelson-Morley apparatus shows the arrangement of the flight paths of the two perpendicular light beams. By using four mirrors instead of one at each end of both flight paths the beams could be reflected back and forth several times; this increased the effective length of the arms from less than four to more than 36 feet, making apparatus 10 times more sensitive than earlier versions. (From "The Michelson-Morley Experiment" by Dr. R.S. Shankland. Copyright (November, 1964) by Scientific American, Inc. All rights reserved.)

Labels in figure: MIRRORS, MIRRORS, LIGHT SOURCE, SILVERED GLASS PLATE, UNSILVERED GLASS PLATE, TELESCOPE, MIRRORS, ADJUSTABLE MIRROR, MIRRORS

producing more intense light (Fig. 2-14). If two light waves are out of step, crests and troughs will cancel each other producing no light (Fig. 2-15). Since the two light beams in the Michelson-Morley apparatus were expected to arrive at slightly different times, they were expected to be slightly out of step producing a fringe pattern—a combination of light and no light. Measurements of the fringe pattern reveal the degree to which the waves are out of step, and further calculations lead to the difference in time for the two light beams to make their respective journeys. But since the apparatus revealed only the difference in time for the two journeys, and not the actual times, Michelson and Morley had to use a more complicated version of the formula we used in

order to calculate the speed of the ether wind, or, its equivalent, the speed of the Earth moving through a stationary ether.

Since the Earth describes an elliptical path as it travels around the Sun, it was impossible to know when one beam of light would be traveling parallel to the ether stream while the other traveled perpendicular to the stream. Consequently, the entire apparatus was mounted on a large sandstone slab, which, in turn floated on mercury. This design helped to dampen vibrations while allowing the apparatus to be turned like a large lazy susan. By making measurements in several different directions, Michelson and Morley could be certain that they would eventually find the "right" direction.

As one might expect, considerable attention was paid to accuracy. For this reason, more than two mirrors were used in the actual apparatus as shown in Figs. 2-16 and 2-17. The two perpendicular light beams were in fact reflected back and forth several times before reaching the telescope. The increased travel distance had the effect of increasing the accuracy of the experiment to the point where the results could be relied upon with substantial confidence.

The stage was finally set. The crucial measurements were made between the 8th and 12th of July, 1887, and much to the chagrin of Michelson and Morley, and the consternation of the entire scientific world, the experiment failed! Regardless of how many times the experiment was repeated, and regardless of how the experimental apparatus was positioned, the expected fringe pattern failed to appear. The time required for the two perpendicular light beams to make their respective round trip journeys was exactly the same. Despite the many years of effort, and the considerable attention paid to detail, man's hope of ever detecting the ether, much less the "absolute motion" of the Earth, vanished forever.

# Insurmountable Difficulties

WHY DID THE MICHELSON-MORLEY EXPERI-
ment fail? Several people came forward to offer explanations. At first, Michelson suggested an idea that is perhaps best understood by imagining that you are standing on a railroad flat car on a day when there is no breeze whatsoever. If the flat car starts moving at 20 miles per hour while you are facing the back of the train, you will feel the wind hitting your back at 20 miles per hour (Fig. 3-1). But if you should start running at 20 miles per hour towards the back of the train, you will no longer feel the wind hitting your back because, relative to the wind, you are not moving at all (Fig. 3-2).

If the entire solar system moved the way the flat car moved, the Earth would feel ether wind hitting it from the "rear". But the Earth, in traveling around the Sun, moves "forward"; and perhaps, suggested Michelson, it moves "forward" at the exact same velocity as the solar system moves "backward". Under these circumstances, the Earth would not be moving relative to the ether wind (Figs. 3-3 and 3-4). To test whether or not this idea was true, Michelson and Morley repeated their experiment six months later. They reasoned that after six months, the Earth, in its journey around the Sun, would be traveling in the opposite direction. The ether wind would then be hitting the "front" of the Earth, and this coupled with the motion of the entire solar system would leave no doubt that the Earth was moving relative to the ether wind. As a matter of fact, just to be doubly certain, Michelson and Morley actually repeated their experiment at three month intervals. Each time, however, the experiment failed.

Michelson then suggested that perhaps ether piles up in front of the Earth the way water would pile up in front of a stone skimming across the surface of a lake (Fig. 3-5). Since the ether which had piled up in front of the Earth would not be moving relative to the Earth, it would be impossible to detect an ether wind. However, several experiments, including one by Michelson, showed that this was simply not true.

Since 1887, the Michelson-Morley experiment has been repeated many times, by many different scientists. Some used far more accurate equipment

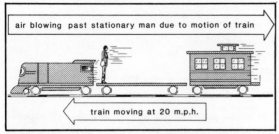

Fig. 3-1. On a day with no breeze, a man standing on a moving train will feel the wind hitting his back.

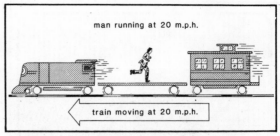

Fig. 3-2. If the man runs as fast as the train moves, he will no longer feel the wind.

than Michelson and Morley; but, needless to say, all obtained the same dismal results.

In 1892, George F. FitzGerald, an Irish physicist, suggested that the Michelson-Morley apparatus itself contracted in the direction of motion (Fig. 3-6). In other words, the distance between the beam splitter and the mirror located upstream, or parallel to the ether current, became just a tiny bit shorter (Fig. 3-7). Furthermore, the decrease in distance was exactly the amount necessary to permit the two light beams to complete their respective journeys in the exact same period of time. But then, you might ask, why not simply measure the

new shortened distance? The answer is that any measuring tape stretched from the beam splitter to the upstream mirror would also shrink, because it would share the same direction of motion as the apparatus itself.

In 1895, the Dutch physicist, Hendrik A. Lorentz independently concluded that the Michelson-Morley apparatus contracted in the direction of motion, and wrote four equations to express the idea in mathematical terms (Fig. 3-8). The first equation can be used to compute the new shortened distance between the beam splitter and the mirror located upstream:

$$
\begin{pmatrix} \text{New (shortened)} \\ \text{distance between} \\ \text{beam splitter and} \\ \text{upstream mirror} \end{pmatrix} = \sqrt{1 - \frac{\left( \begin{array}{c} \text{Velocity of Michelson-Morley} \\ \text{apparatus relative to ether} \end{array} \right)^2}{(\text{Velocity of light})^2}} \begin{pmatrix} \text{Measured distance} \\ \text{between beam} \\ \text{splitter and} \\ \text{upstream mirror} \end{pmatrix}
$$

EQUATION 1

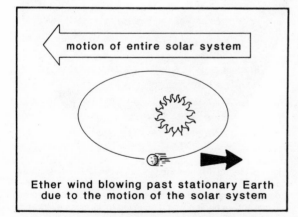

Fig. 3-3. If the entire solar system moved through a stationary ether, the Earth would "feel" ether wind hitting it from the "rear."

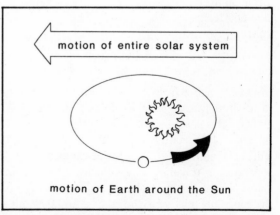

Fig. 3-4. If the Earth moved "forward" as fast as the entire solar system moved "backward," the Earth would no longer "feel" the ether wind.

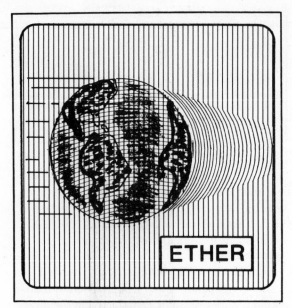

Fig. 3-5. If ether piled up in front of the Earth, it would be stationary relative to the Earth and therefore impossible to detect by the Michelson-Morley apparatus.

Suppose the measured distance between the beam splitter and upstream mirror is 2 feet, and the velocity of the Michelson-Morley apparatus relative to the ether is 10,000 miles per second. The equation tells us that the new shortened distance is:

$$\text{New (shortened) distance between beam splitter and upstream mirror} = \sqrt{1 - \frac{(10,000)^2}{(186,000)^2}} \quad (2)$$

$$= 1.997 \text{ feet}$$

Obviously, a drop from 2 feet to 1.997 feet amounts to hardly any change at all despite the fact that 10,000 miles per second is hardly a slow speed. But the truth of the matter is that the change in length will not be noticeable unless the Michelson-Morley apparatus is traveling at a velocity close to the speed of light. For example, suppose the velocity of the apparatus relative to the ether is 180,000 miles per second. The new shortened distance is then:

$$\text{New (shortened) distance between beam splitter and upstream mirror} = \sqrt{1 - \frac{(180,000)^2}{(186,000)^2}} \quad (2)$$

$$= 0.504 \text{ feet}$$

If we could perform the Michelson-Morley experiment while traveling at 180,000 miles per second relative to the ether, we could expect to see the distance between the beam splitter and upstream mirror shrink to 0.504 feet or very close to six inches. Unfortunately, if we tried to measure this new distance, we would find that it still measures 2 feet because it would be necessary to take our

Fig. 3-6. George Francis FitzGerald. A physicist whose many accomplishments included the suggestion that electromagnetic waves could be produced by an oscillating electric current—a finding which soon led to wireless telegraphy (courtesy of Brown Brothers).

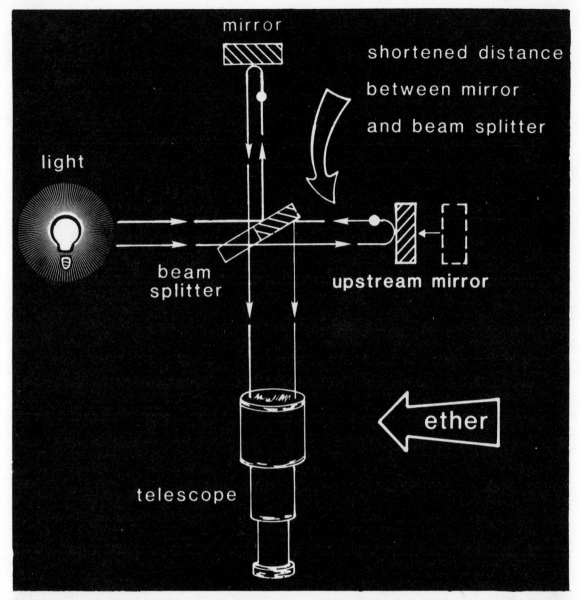

mirror

shortened distance
between mirror
and beam splitter

light

beam
splitter

upstream mirror

ether

telescope

Fig. 3-7. FitzGerald suggested that perhaps the distance between the beam splitter and upstream mirror became a tiny bit shorter. Under these circumstances the two perpendicular light beams could complete their respective journeys in the exact same period of time.

measurement with a shrunken ruler. Such were the strange predictions of the first equation of Hendrik A. Lorentz.

All this shrinkage, of course, takes place only in the direction of motion. In his second equation,

Lorentz said that no shrinkage takes place between the beam splitter and the mirror located across the ether stream; in other words, no shrinkage takes place in a direction perpendicular to the direction of motion.

Fig. 3-8. Hendrik Antoon Lorentz. Although he wrote the equations which appeared in the Special Theory of Relativity, he was unable to provide the correct theory. Nevertheless, for his many accomplishments in physics he earned the Nobel Prize in 1902 (courtesy of Brown Brothers).

New distance between beam splitter and cross-stream mirror = Measured distance between beam splitter and cross-stream mirror

### EQUATION 2

In his third equation, Lorentz said that no shrinkage takes place between the beam splitter and anything that might be located above or below the beam splitter. Again, Lorentz is simply saying that no shrinkage takes place in a direction perpendicular to the direction of motion.

New distance between beam splitter and anything located above or below beam splitter = Measured distance between beam splitter and anything located above or below beam splitter

### EQUATION 3

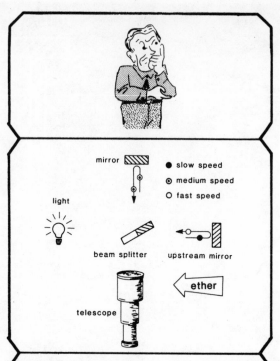

In the Michelson-Morley Experiment, it was assumed that a photon traveling against the ether current (to the upstream mirror) and back would take a tiny fraction of a second longer to complete the journey than a photon traveling across the ether current and back. While traveling to the upstream mirror, a photon would be subject to the retarding effects of the current and move slowly, but after turning around it would benefit by the motion of the current and move quickly. In contrast, the photon traveling across the ether current and back would complete both legs of the journey at some intermediate speed.

Fig. 3-9.

In summary, the first three equations written by Lorentz state that the shrinkage experienced by the Michelson-Morley apparatus occurs in only one of its three dimensions. Furthermore, the shrinkage is just sufficient to permit the two light beams to complete their respective round-trip journeys in the exact same period of time—a fact, which, as we shall see, raised a far more serious problem for Lorentz.

Although convinced that the distance between the beam splitter and the upstream mirror had shrunk, Lorentz still maintained that the ether

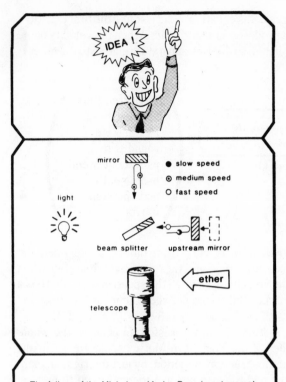

The failure of the Michelson-Morley Experiment prompted Fitzgerald and Lorentz to suggest that perhaps the distance between the beam splitter and the upstream mirror became just a tiny bit shorter. With less distance to cover, a photon traveling against the ether current (to the upstream mirror) and back would take the same time as a photon traveling across the ether current and back. It was still assumed, however, that a photon would move slowly towards the upstream mirror and quickly away from it, while the other photon moved back and forth across the ether current at some intermediate speed.

Fig. 3-10.

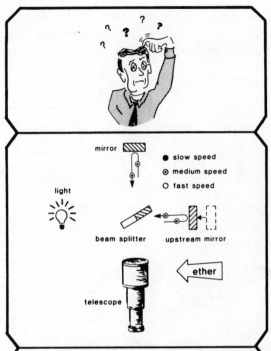

Suspecting that no experiment could ever be devised to show that the speed of light varies with direction, and still maintaining that the distance between the beam splitter and upstream mirror became a tiny bit shorter, Lorentz tried to explain how two photons traveling at the same speed could cover different distances in the same time. He was forced to create the concept of "artificial" time - the time required for a photon to complete the round-trip journey between the beam splitter and upstream mirror. Using an equation which he developed to calculate this time, Lorentz showed that the photon traveling between the beam splitter and the upstream mirror took less time than the photon traveling across the ether current and back. Since this made no sense at all, Lorentz concluded that "artificial" time was a mathematical necessity with no relation to reality.

Fig. 3-11

would force light to travel more slowly towards the upstream mirror than away from it. Realizing, however, that the Michelson-Morley experiment had failed to demonstrate that the speed of light varies with direction, Lorentz began to question if any experiment could ever be devised to show that the speed of light varies with direction. Concluding that the answer was definitely no, Lorentz found that he had to explain how the two perpendicular light beams, traveling at the same speed in every direction, could each cover different distances in the exact same period of time. This was like trying to explain how two cars, each traveling at a constant 55 miles per hour, could manage to cover different distances in exactly one hour. To resolve this dilemma, Lorentz came to the uncomfortable conclusion that each light beam completed its round trip journey in a different period of time! The light beam traveling up the ether stream and back took less time than the light beam traveling across the ether stream and back. Since it seemed certain that scientists would never be able to measure this shorter period of time, Lorentz called it an "artificial" time—a time with no real physical meaning contrived only to explain the results of the Michelson-Morley experiment. Lorentz embodied this concept in his fourth equation:

Since this equation looks just like the first equation, it produces similar results. When the velocity of the Michelson-Morley apparatus relative to the ether is

"Artificial" time for light beam to complete round trip journey between beam splitter and upstream mirror $= \sqrt{1 - \dfrac{\left(\begin{array}{c}\text{Velocity of Michelson-Morley} \\ \text{apparatus relative to ether}\end{array}\right)^2}{(\text{Velocity of light})^2}} \left(\begin{array}{l}\text{Measured time} \\ \text{for light beam} \\ \text{to complete} \\ \text{round trip} \\ \text{journey between} \\ \text{beam splitter} \\ \text{and upstream} \\ \text{mirror}\end{array}\right)$

EQUATION 4

"DOC" QUACKER'S
'Artificial'
Time Tonic

Fig. 3-12.

low, there is very little difference between the measured time and the "artificial" time. But when the velocity is close to the speed of light, there is a marked difference between the measured time and "artificial" time (Figs. 3-9, 3-10, and 3-11).

And so matters rested. In general, the whole idea of the Michelson-Morley apparatus shrinking in the direction of motion, by just the right amount, seemed too artificial and too contrived to be real. The concept of "artificial" time seemed so strange that even its originator expressed misgivings while advancing the idea (Fig. 3-12).

Matters had clearly gone from bad to worse. But if the notion of the ether was to be maintained, there seemed to be no other explanation. The only other alternative was to discard the notion of the ether altogether. But then what would replace it? In 1905, a 26 year old Swiss patent clerk came up with the answer.

**Chapter 4**

# The Genius

Ｈ IS NAME WAS ALBERT EINSTEIN. HE WAS BORN on March 14, 1879 in Ulm, Germany to a family that had never distinguished itself academically or intellectually. A year after his birth, the family moved to Munich. It was here that his father, Hermann Einstein, and his uncle, Jakob Einstein, set up an electrochemical works. A year after the move, Albert's sister, Maja, was born. They were the only children of Hermann and Pauline (nee Koch) Einstein (Figs. 4-1 and 4-2).

As a child, Albert showed no particular signs of genius. He learned to speak at a later age than most children, and still hesitated to respond to questions at nine years of age. His mother came from a fairly well established family, and instilled in her son a love for classical music. At the age of six he began to take violin lessons, and became quite an accomplished amateur player as an adult. At the age of seven, Albert's Uncle Jakob, an engineer, started teaching him algebra. And at the age of thirteen, a young medical student and family friend, Max Talmey, introduced him to physics and philosophy.

Talmey later wrote that young Albert quickly absorbed his readings in philosophy, and displayed a keen interest in the philosophical writings of Immanuel Kant. But despite these interests, Albert's performance in school left much to be desired. His deep resentment of the formal, rigid methods common to the German schools at that time was plainly evident in his attitude (Fig. 4-3).

At the age of sixteen, his father's business failed whereupon the family moved to Italy. Albert, however, was left in Germany to finish school alone—a situation he found intolerable. After six months, he decided that he had had enough of both the German high school (Luitpold Gymnasium) and living alone. He went to a doctor and obtained a note saying that he had suffered a nervous breakdown. He then went to his math teacher and obtained a note indicating that his proficiency in mathematics provided sufficient qualification for entrance into a university despite the fact that he had no high school diploma. With these two notes in hand, Albert was about to leave school when he was notified

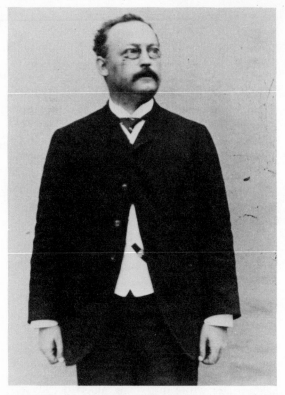

Fig. 4-1. Einstein's parents, Hermann and Pauline Einstein. A devoted couple, they provided a happy and comfortable home for their two children. Hermann was a jovial, optimistic man constantly plagued with business difficulties. Pauline was artistically inclined and always eager to play the piano (courtesy of Hebrew University of Jerusalem Source: American Institute of Physics).

that he had been dismissed. He was informed that his presence in the classroom disrupted the other students.

Albert went to Italy, joining his family in Milan, and during the year that followed did essentially nothing with the exception of some mountain climbing and a little self-tutoring in mathematics. At the end of the year, his father's business failed again. Realizing that the time had come for his son to begin supporting himself, Albert's father encouraged him to apply to the Zurich Polytechnic Institute and major in electrical engineering. Albert took the entrance examination and failed. He did well enough on the math portion, however, to be told that he could enter the university without taking another examination if he agreed to get a high school diploma. He subsequently enrolled in a high school in Aarau, Switzerland, a town several miles

west of Zurich, and earned his high school diploma by the end of the year.

In 1896, he entered the Zurich Polytechnic where he spent four undistinguished years. He managed to get through college largely with the help of his friends. One of these friends, Marcel Grossman, a mathematician who would later work with Einstein, took very detailed, copious notes. Einstein studied from these notes in order to pass several important examinations. He graduated in August, 1900 with an overall gradepoint average of 4.91 out of 6.00.

A typical graduate of the Polytechnic, like Einstein, usually managed to find a job as an assistant to a professor at some university, if not the Polytechnic itself. And although many of his classmates did, indeed, acquire such positions, Einstein was not to be as fortunate. The reason has

Fig. 4-2. Albert at the age of fourteen and his sister Maja at the age of twelve. The two remained lifelong companions. When she died in December 1951, Einstein wrote " . . . I miss her more than one can imagine" (courtesy of Hebrew University of Jerusalem Source: American Institute of Physics).

different positions including teacher and tutor. Also, in December, 1900, he published his first paper, "Deductions from the Phenomena of Capillarity" in the Annalen der Physik (Annals of Physics). But for the most part, these were two frustrating, difficult years for the young Einstein. An allowance from relatives which had provided support through college was no longer available. Efforts to acquire steady work met with little or no success, and periods of unemployment were not infrequent.

Once again, Marcel Grossman came to the rescue. Grossman, feeling sorry for his friend, spoke to his father who happened to be a friend of Friedrich Haller, the director of the Swiss Patent Office. Herr Grossman managed to secure a promise from Haller that he would interview young Einstein when a position became available. In time, Einstein was interviewed and offered a job which he eagerly accepted. On June 23, 1902, he assumed the position of a patent examiner with the title of Technical Expert Third Class.

Einstein kept his job as patent examiner for seven years—years which he would later refer to as some of the happiest of his life (Fig. 4-4). On Jan-

more to do with Einstein's personality in those years than with his intellect. Einstein's public image as the good-natured gentle, professor lies in stark contrast to the college aged youth. After graduation, one professor said to him, "You're a clever fellow! But you have one fault. You won't let anyone tell you a thing . . ." Years later, Einstein would describe himself during those days as "untidy and a daydreamer . . . aloof and discontented, not very popular." Having been turned down by every professor at the Polytechnic, he began to look elsewhere.

During the next two years, he held a number of

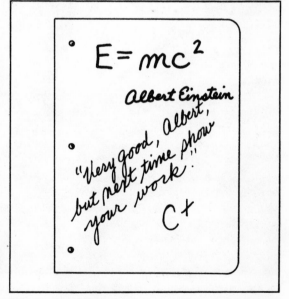

$$E = mc^2$$

Albert Einstein

"Very good, Albert, but next time show your work."

C+

Fig. 4-3. Contributed by Warren K. Heyer. Reprinted with permission from the March 1982 Reader's Digest.

Fig. 4-4. Albert Einstein at work in the Patent Office. It has been said that he thought more like an artist than a scientist. Rather than describe a good theory as correct or exact, he would say it was beautiful (courtesy of Hebrew University of Jerusalem Source: American Institute of Physics).

uary 6, 1903 he married Mileva Maric, a classmate at the Zurich Polytechnic Institute. Hans Albert, their first son, was born later than year (Fig. 4-5). During this period, he lived the life of a poor civil servant. After working eight hours a day, he would come home and quickly bury himself in physics. Friends who visited in those years said that they would usually find him surrounded by a sea of books and papers, intensely concentrating on something he was reading; and while the air in the room reeked from the smell of a stinking cigar, he would sit absent-mindedly rocking his son in a bassinet. He continued to publish papers, and in 1905 sent five papers to the Annalen der Physik which collectively revolutionized both physics and philosophy.

With his first paper in 1905, "A New Determination of Molecular Dimensions," he obtained his Ph.D. from the University of Zurich. It was the least distinguished of the five papers published that year. In his second paper, "On the Motion of Small Parti-cles Suspended in a Stationary Liquid According to the Molecular Kinetic Theory of Induction", he confirmed the existence of molecules. Even as late as 1905, there were physicists who doubted their existence. In his third paper, "On a Heuristic Viewpoint Concerning the Production and Transformation of Light", he proposed that light consisted of bundles of energy called quanta—later to be called photons. This paper laid the theoretical foundation for what would eventually become the photoelectric cell. His fourth paper that year, "Zur Elektrodynamik bewegter Korper" ("On the Electrodynamics of Moving Bodies") contained the principles of the Special Theory of Relativity. And in his fifth paper, "Ist die Tragheit sines Korpers von seinem Energieinhalt abhangig?" ("Does the Inertia of a Body Depend Upon Its Energy Content?") he wrote his famous formula, $E = mc^2$. Although there were several physicists who immediately perceived the significance of these pap-

Fig. 4-5. Einstein's first wife Mileva Maric with sons Hans Albert (right) and Eduard. Following their divorce, Mileva retained the name Einstein, and lived another quarter of a century caring for her younger son who was diagnosed as a schizophrenic. Hans Albert eventually settled in the United States and became a professor of hydraulic engineering at Berkeley (courtesy of Hebrew University of Jerusalem Source: American Institute of Physics).

ers, most were slow to absorb these revolutionary ideas. Needless to say, the papers passed entirely unnoticed at the patent office. On April 1, 1906, Albert Einstein was promoted from Technical Expert Third Class to Technical Expert Second Class*!

In 1907, Alfred Kleiner, a professor at the University of Zurich, suggested to Einstein that he apply for the position of privadozent in the University of Berne. A privadozent was a lecturer who worked for a nominal sum, and a position which was required in Europe before a person could be appointed to the faculty of some university. Along with his application for the job, Einstein submitted his paper on relativity. He was turned down with the excuse that his paper was incomprehensible. It was not until Kleiner interceded on his behalf that he was appointed privadozent in 1908. In the spring of 1909, after overcoming additional resistance, Einstein was appointed associate professor at Zurich University.

With his ideas now beginning to spread, Einstein was rapidly becoming one of the more important physicists in Europe. In 1911, he became a full professor at the German University in Prague. But always anxious to be with people who might help him with his work, Einstein moved back to Zurich in the winter of 1912 to assume a professorship at his alma mater, the Zurich Polytechnic. Here he could work with his old friend Marcel Grossman on a new theory of gravity which he called the General Theory of Relativity. The prodigious effort required for this work is perhaps best revealed in a letter to a colleague:

> "I am now exclusively occupied with the problem of gravitation and hope to overcome all of the difficulties. Never in life have I been quite so tormented. A great respect for mathematics has been instilled in me, the subtler aspects of which, in my

stupidity, I regarded until now as pure luxury. Compared with the problem of gravitation, the original problem of the theory of relativity is child's play."

In 1913, Einstein and Grossman published a paper titled, "Outline of a General Theory of Relativity and a Theory of Gravitation". Although the paper showed that progress had been made, work on the theory of gravity was not nearly complete. Einstein labored another three years before finally publishing, "Die Grundlagen der allgemeinen Relativitatstheorie" (The Foundation of the General Theory of Relativity"), a paper which, in time, would make him the world's most famous scientist.

In the summer of 1913, Einstein was invited to

Fig. 4-6. Elsa Lowenthal—Einstein's second wife. Maternal, protective and undemanding, she was in many ways an ideal wife for the world famous Einstein. She once said, "All I can do is to look after his outside affairs, take business matters off his shoulders, and take care that he is not interrupted in his work" (courtesy of the Mansell Collection).

---

*Of course, Einstein's interest in physics was of no concern to Friedrich Haller. However, by this time, Haller had come to value Einstein as an employee, and expressed regret when he left the patent office in 1909.

Fig. 4-7. Albert Einstein—(circa 1920) Once asked for the whereabouts of his laboratory he removed a fountain pen from his breast pocket and with a smile said, 'Here' (courtesy of Brown Brothers).

assume the position of Director of the newly formed Kaiser Wilhelm Institute, a position in which he could teach or do research as he pleased. Despite the fact that he would have to move back to Germany, the offer was too good to turn down. The salary, to begin with, was excellent; and although Einstein never cared about money, he now had two sons to raise (his second son, Eduard, was born in 1910). But perhaps the most enticing aspect of the offer was the fact that he would not have to lecture, and could devote all of his energies to the General Theory of Relativity. Nevertheless, his distaste for Germany became evident once again when it came to dealing with the question of citizenship. After leaving Germany at the age of 16, Einstein had renounced his German citizenship, and, in time, became a Swiss citizen. Now, he agreed to return to Germany only under the condition that he would not be required to become a German citizen again. With this understood, he assumed his new position in April, 1914.

Einstein's marriage to Mileva had never been a particularly happy union, and with the move to Germany, the relationship grew worse. In the summer of 1914, Mileva and the boys returned to Zurich for a vacation while Einstein remained in Berlin. When the war broke out in August, it was decided that the family should remain separated; a separation which finally led to divorce. Although Einstein was happy to be free of Mileva, he had no desire to see the boys suffer from the divorce. Getting money out of Germany and into neutral Switzerland for the support of the boys became one of Einstein's chief preoccupations during the war. The money which accompanied his Nobel Prize in Physics (1921) was also turned over to Mileva for the support of her and the boys.

In 1919, Einstein married again, this time to Elsa Lowenthal, the widowed daughter of his father's cousin and the mother of two girls (Fig. 4-6). It was also in this year the he became a world figure (Fig. 4-7). But for the moment, let us not get ahead of our story. Having approached that great monument called the Theory of Relativity, let us now attempt to scale its heady heights.

# The Special Theory of Relativity

**Chapter 5**

E INSTEIN'S GREAT WORK WHICH APPEARED in the paper, "On the Electrodynamics of Moving Bodies", resulted in a complete revision of the very foundations of physics. What do we mean by such a grand statement? To find out, we must first ask ourselves, "What do physicists do?" The answer is that they do what in fact all scientists do—they measure things. Whatever they measure, be it velocity, acceleration, force, or anything else, can usually all be reduced to units of length, mass, and time (Fig. 5-1). This is why we refer to the metric system as the CGS system, or centimeter-gram-second system. Let us consider a few examples to illustrate why most measurements can almost always be reduced to units of length, mass, and time.

It is not at all unusual for us to drive around the neighborhood at a velocity* of 25 miles per hour.

---

*For the sake of simplicity, the terms speed and velocity are used interchangeably throughout this book. In fact, physicists have very specific definitions for these terms. Roughly speaking, speed is distance covered per unit of time, velocity is speed in a specified direction.

Velocity, therefore, can obviously be broken down into units of length and time: miles and hours. If we accelerate very, very slowly from 25 miles per hour to 55 miles per hour in a time period of one hour, our acceleration is 30 miles per hour per hour. In other words, each hour we increase our velocity by 30 miles per hour. With acceleration given in units of miles per hour per hour, we can easily see that acceleration, like velocity, can also be broken down into units of length and time.

When we throw a baseball, we swing our arm forward and let the ball fly. In other words, we actually accelerate the ball from a velocity of 0 miles per hour to, let us say, 25 miles per hour at the instant of release. In the 17th century, Isaac Newton determined that the force we exert when we throw a ball can be calculated by multiplying the acceleration of the ball by its mass.

$$Force = Mass \times Acceleration$$

We have already learned that acceleration can be

Fig. 5-1. Whatever physicists measure can usually be separated into units of length, mass, and time.

broken down into units of length and time. Force, therefore, a combination of mass and acceleration, can ultimately be broken down into units of mass, length, and time.

Since mass, length, and time are the fundamental units of measure, it should come as no surprise that any change in our ideas concerning these units amounts to a change in the very foundations of physics. In the Special Theory of Relativity, Einstein succeeded in changing the way we think about length, mass, and time, and that is why we say that he changed the very foundations of physics.

In creating the Special Theory of Relativity, Einstein began by deciding that there were two things going on in the universe which had never been recognized by scientists, but which were so important that they could now be regarded as irrefutable facts of life. The first fact states:

> It is impossible to detect the motion of the Earth, or any other heavenly body, relative to an ether assumed to be standing perfectly still in the universe.

35

Consequently, it is impossible to know if any heavenly body is truly standing still or moving in the universe.

The second fact states:

The speed of light is the same regardless of whether the light source is moving or not, and regardless of whether the observer is moving or not.

In his statement of the first fact, Einstein said that the Michelson-Morley Experiment failed to detect the absolute motion of the Earth because no experiment, however sophisticated, and however accurately devised, could ever successfully detect the absolute motion of the Earth. The universe is simply so designed that man would forever be unable to know how fast the Earth is moving relative to all creation. Furthermore, since we can never detect the absolute motion of the Earth, we will never know if an ether exists or not. And, as we will soon see, despite all of the effort expended in trying to detect the ether, it becomes unnecessary to suppose that it even exists.

In his statement of the second fact, Einstein allowed himself to be guided by his remarkable intuition—an intuition which the world would later recognize as genius. To appreciate just how remarkable his intuition really was, we will examine his second fact in considerable detail.

Returning to the railroad flat car, imagine that a friend is standing at the end of the flat car next to the caboose. The train is standing perfectly still, and our friend, who is facing the front of the train, throws a ball toward the engine at 30 miles per hour (Fig. 5-2). Since we are standing on the ground next to the railroad track, we see the ball cross our line of vision at 30 miles per hour. Of course, our friend also sees the ball leave his hand at 30 miles per hour. Now, if the train starts moving forward at 20 miles per hour, and our friend throws the ball forward at 30 miles per hour; he will see the ball leave his hand at 30 miles per hour, but we will see the ball cross our line of vision at 50 miles per hour (Fig. 5-3). Because of the motion of the train, the ball had a speed of 20 miles per hour before it even left our friend's hand. When our friend threw the ball, it picked up an additional 30 miles per hour, and as a result, we see it cross our line of vision at 50 miles per hour. So far, all of this conforms to common everyday experience, so we have no reason to feel uncomfortable.

The speed of sound depends only on the medium in which it travels, and that usually means air. Sound travels at approximately 750 miles per

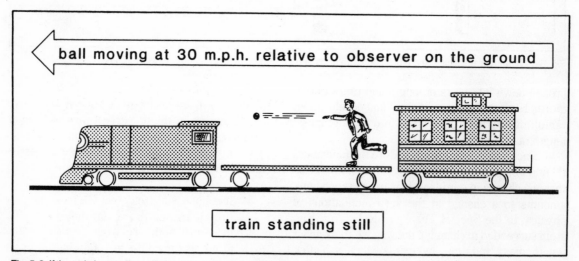

Fig. 5-2. If the train is standing still, the man on the train and an observer on the ground will both see the ball move at 30 m.p.h.

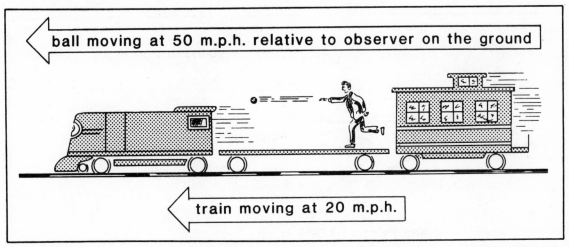

Fig. 5-3. If the train is moving, the man on the train will see the ball move at 30 m.p.h. and an observer on the ground will see the ball move at 50 m.p.h.

hour in air, and it doesn't matter how fast the source of the sound is moving. The pilot of a jet plane traveling at 600 miles per hour, would find that the sound of his engine moves ahead of his plane at only 150 miles per hour (Fig. 5-4). If our pilot were traveling at 1000 miles per hour, he would actually fly faster than the sound of his engine (Fig. 5-5). This phenomenon was dramatically illustrated during World War II when the Nazis attacked London with V-2 rockets. These rockets flew faster

than sound, and so the sound of the arriving rocket would be heard only after the rocket had exploded.

Now, according to Einstein, a photon of light does not behave like either the ball on the railroad flat car, or the sound emanating from a jet engine. Imagine a flying saucer, equipped with landing lights, and capable of flying at speeds close to the speed of light. If the saucer sits motionless on the landing pad, and the pilot turns on the landing lights, the pilot will see a photon of light move away from

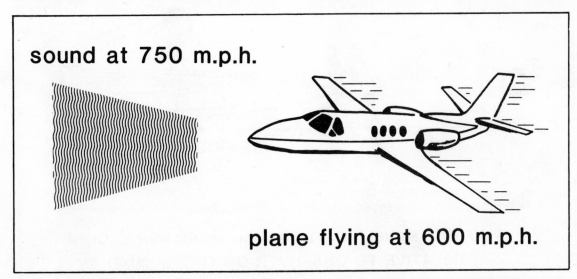

Fig. 5-4. If a plane is traveling at 600 m.p.h., the sound of its engines will travel 150 m.p.h. faster than the plane.

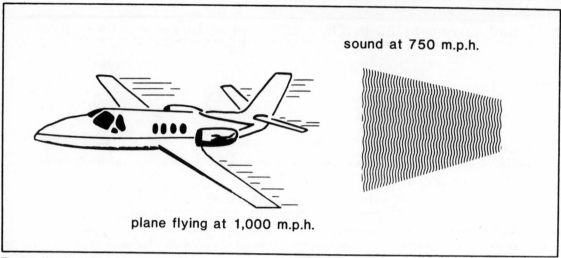

Fig. 5-5. If a plane is traveling at 1000 m.p.h., the sound of its engines will travel 250 m.p.h. slower than the plane.

the saucer at 186,000 miles per second (Fig. 5-6). If we stand next to the saucer, we too will see a photon move away from the saucer at 186,000 miles per second. But if the saucer is flying across our line of vision at 150,000 miles per second, and the pilot turns on the landing lights, he will see a photon of light move away from the saucer at 186,000 miles per second; and according to Einstein, we too will see the photon move away from the saucer at 186,000 miles per second! We will not see the photon move away from the saucer at 336,000 miles per second (186,000 plus 150,000). Of course, this

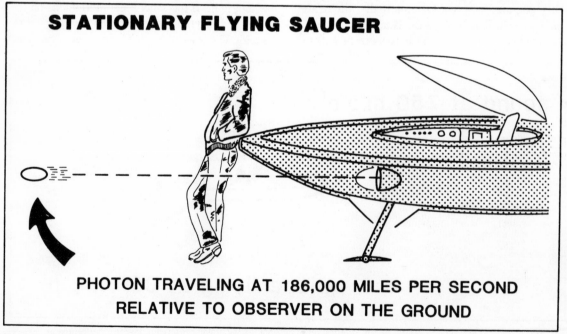

Fig. 5-6. A photon leaving a stationary light is observed to travel at 186,000 miles per second.

PHOTON TRAVELING

AT 186,000 MILES PER SECOND

RELATIVE TO OBSERVER ON THE GROUND

Fig. 5-7. A photon leaving a moving light is also observed to travel at 186,000 miles per second.

completely contradicts common sense (Fig. 5-7). It is certainly no surprise that the pilot sees the photon moving away from him at 186,000 miles per second regardless of whether or not the saucer is in motion. After all, our friend on the railroad flat car saw the ball move away from him at 30 miles per hour regardless of whether or not the train was in motion. However, we saw the ball cross our line of vision at 30 miles per hour when the train was standing still, and at 50 miles per hour when the train was moving. Now it turns out that we will see the photon move away from the flying saucer at 186,000 miles per second regardless of whether the saucer is moving or not. But after considering this matter for over 10 years, Einstein finally concluded that this is precisely what we would see.

What is most remarkable is that Einstein had no experimental evidence to guide him to this bizarre conclusion. It was essentially an exercise in intuition. Only in 1913, eight years after its publication, were astronomers able to show that the speed of light was the same regardless of the motion of the source. In their observations of binary stars, two stars which revolve around each other, astronomers observed that the speed of light from both stars was the same despite the fact that one star was traveling towards the Earth, while, at the same time, the other was traveling away from the Earth (Fig. 5-8).

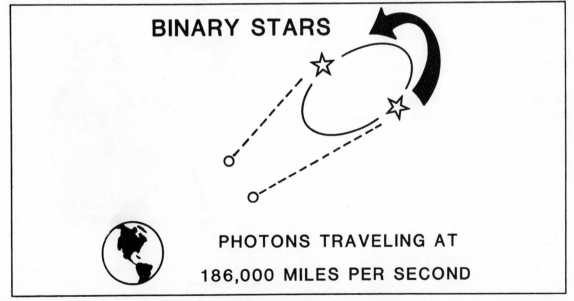

BINARY STARS

PHOTONS TRAVELING AT
186,000 MILES PER SECOND

Fig. 5-8. In a binary star system, one star travels away from the Earth while the other travels towards the Earth. Nevertheless, photons emanating from both stars approach the Earth at 186,000 miles per second.

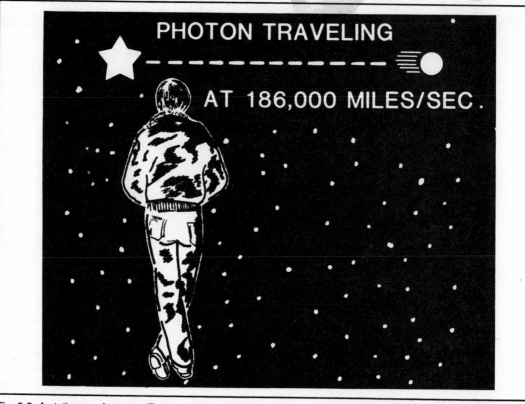

Fig. 5-9. A stationary observer will see a photon of light travel away from a star at 186,000 miles per second.

Just as the speed of light will be the same regardless of the motion of the source, so, too, will it be the same regardless of the motion of the observer. If you were to watch the light from a star cross your line of vision, you would see it moving at 186,000 miles per second (Fig. 5-9). If you were a passenger in the flying saucer traveling past the star at 150,000 miles per second, you could look out the window and still see the light from the star crossing your line of vision at 186,000 miles per second (Fig. 5-10). Obviously, the point of all this is that the speed of light is constant; so constant, in fact, that it doesn't matter if the light source or observer is moving or not.

Starting with his two facts about the universe which had been previously disregarded by scientists, Einstein was now in a position to alter our notions of mass, length, and time. However, as we have just seen, one of these facts completely con-

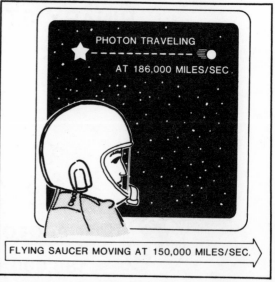

Fig. 5-10. A moving observer will also see a photon of light travel away from a star at 186,000 miles per second.

tradicted common sense. Before 1905, scientists made statements which at least "felt" right. For example, Newton's third law of motion, for every action there is an equal and opposite reaction, was something that could be observed everyday at any billiard table. But from now on, man would have to deal with a universe in which more than a sharp eye and keen insight were required to unlock the secrets of nature. Increasingly, the rules of logic and mathematics would have to be relied upon in order to successfully navigate this new abstract universe which Einstein had begun to map.

Having started with the abstract, it is not surprising that Einstein ended with the abstract. The conclusions reached in the Special Theory of Relativity make as little common sense as the behavior

of a photon of light. Strict adherence to logic and mathematics is the only way to reach these conclusions. It is no wonder that the Theory of Relativity was so difficult to understand when it was first published. Never before had scientists or the public been asked to apply so much discipline to their thinking.

We begin our journey through the abstract by first examining length and time. Imagine a man standing on a special kind of Sun. This Sun operates like a large light bulb. If the man presses a switch, the Sun will turn on just like any average light bulb. Ninety-three million miles away, a man on the Earth is waiting to see the light from this unusual Sun (Fig. 5-11). At this distance, he can expect to see the light approximately eight minutes after the

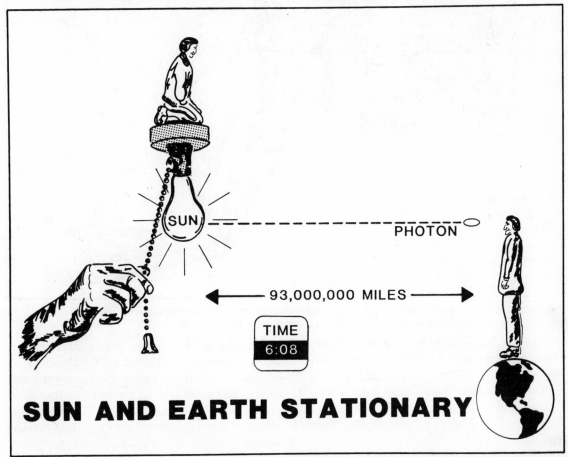

Fig. 5-11. It takes approximately eight minutes for a photon of light to traverse the 93,000,000 miles that separates the Sun from the Earth.

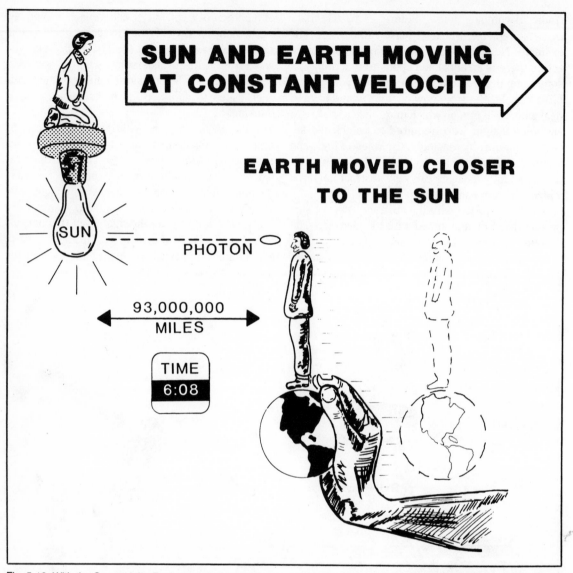

**SUN AND EARTH MOVING AT CONSTANT VELOCITY**

**EARTH MOVED CLOSER TO THE SUN**

SUN

PHOTON

93,000,000 MILES

TIME 6:08

Fig. 5-12. With the Sun and the Earth moving, the photon cannot reach the Earth in eight minutes. One way to remedy the situation is to move the Earth closer to the Sun and call the new shortened distance 93,000,000 miles.

man on the Sun presses the switch. Although we might be hard pressed to find a Sun that operates like a light bulb, it is true that our Sun is 93,000,000 miles away, and that it takes eight minutes for light from the Sun to reach the Earth. Whenever we look at the Sun, we see it as it appeared eight minutes ago.

Suppose now, that this Sun and this Earth were moving through the universe in a straight line from left to right, and at a constant velocity. The distance between them would still be 93,000,000 miles. Furthermore, neither the man on the Sun nor the man on the Earth could detect the fact that they were moving at all. Remember the first fact: it is impossible to know if any heavenly body is truly standing still or moving in the universe. These two men may even go to the trouble of performing some clever, sophisticated experiment like the Michelson-

Morley experiment, but they will find, as indeed Michelson and Morley did, that it is impossible for them to detect their motion in space. As both men look around, they see stars, planets, and galaxies passing them. As we "stand" out in space watching them, they will even see us passing them. But they will never be able to determine if they are passing everything in the universe, or if everything in the universe is passing them. To make matters worse,

we are in the same position they are. We, too, cannot be certain that we are standing still in the universe. The only thing we and our friends can both maintain with certainty is that we are both moving relative to each other. Having decided that they are moving relative to us, let us consider what we might observe.

Suppose that at 6:00, the man on the Sun presses the switch turning the Sun, "on", and a light

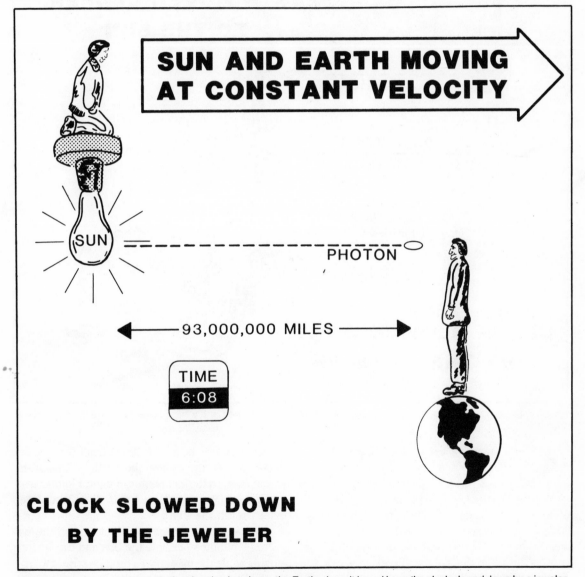

Fig. 5-13. Another way to remedy the situation is to leave the Earth where it is and have the clock slowed down by a jeweler.

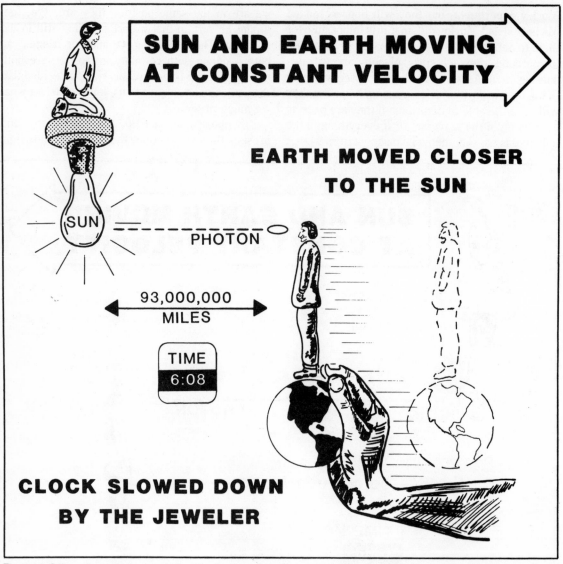

# SUN AND EARTH MOVING AT CONSTANT VELOCITY

### EARTH MOVED CLOSER TO THE SUN

SUN

PHOTON

93,000,000 MILES

TIME 6:08

## CLOCK SLOWED DOWN BY THE JEWELER

Fig. 5-14. Still another way to remedy the situation is to do both—move the Earth a littler closer and have the clock slowed down.

beam starts traveling towards the Earth. With a distance of 93,000,000 miles to travel, a photon of light should reach the Earth at 6:08. As a matter of fact, if this Sun and Earth were standing still in front of us, we could watch the photon take 8 minutes to complete its journey. But the reality is that this Sun and Earth are moving across our line of vision; consequently, as the photon travels towards the Earth we see the Earth moving away from the

photon. At 6:08, the photon has still not reached the Earth. We see the man on the Earth look up at the sky and search for that photon of light. Finally, after let us say another 2 minutes, 6:10, the photon does indeed reach the Earth.

We saw the photon take longer than the expected 8 minutes because the Sun and Earth were moving while the photon was traveling. But the man on the Sun and the man on the Earth have every reason to believe that they are standing still and everything is passing them. They have no way of proving that they are moving. They also know that the photon traveled 93,000,000 miles in 10 minutes and not the expected 8 minutes. They are well aware of the formula:

$$\text{Time} = \frac{\text{Distance}}{\text{Velocity}}$$

and they realize that according to this formula, the photon should have traveled for approximately 8 minutes:

$$\text{Time} = \frac{\text{Distance}}{\text{Velocity}} = \frac{93,000,000 \text{ miles}}{186,000 \frac{\text{miles}}{\text{second}} \times 60 \frac{\text{seconds}}{\text{minute}}}$$

$$\text{Time} = \text{approximately 8 minutes}$$

How do we resolve this dilemma? There are four alternatives that we can examine.

First, our two friends might get themselves a faster light beam, or, to use the words of our story, a faster photon. But that is impossible! Remember the second fact: the speed of light is the same regardless of whether the observer is moving or not. In other words, there simply is no such thing as a faster photon. A photon travels at 186,000 miles per second, and that is all there is to it! As a second possibility, our two friends might elect to move the Earth closer to the Sun, and call the new, smaller distance 93,000,000 miles (Fig. 5-12). Under these circumstances, the photon would, in fact, reach the Earth in 8 minutes after traveling "93,000,000 miles". As a third possibility, they might leave the Earth where it is, and take their clocks to the jeweler and have them slowed down (Fig. 5-13). Under these circumstances, the photon would reach the Earth in "8 minutes" after traveling 93,000,000 miles. The fourth possibility, a combination of the last two, is in fact what really does happen. The Earth is moved a little closer to the Sun, and the clocks are slowed down.* In other words our two friends, who find themselves unable to detect their motion in space (Fact One) and stuck with a photon whose speed cannot be altered (Fact Two), are forced into a situation where they must shorten the distance they call 93,000,000 miles and increase the interval of time they call 8 minutes (Fig. 5-14). To make all of this easier to understand, let us consider the changes to length and time separately, realizing, nevertheless, that both occur together.

---

*In order to avoid the difficult concept of simultaneity which lies at the heart of the Special Theory, the reader is asked to accept the fact that both occur together.

**Chapter 6**

# Length

I N DISCUSSING LENGTH, WE WILL CONTINUE TO use the model of the man on the Sun and the man on the Earth. However, we will now watch from our own Sun-Earth System. When we look up in the sky, we will be able to see our friends on their Sun-Earth System with their Sun that operates like a giant light bulb. And when they look up in the sky, they will see us peering at them from our Sun-Earth System.

We must also establish the fact that a radio signal travels at the speed of light, 186,000 miles per second, because it is an electromagnetic wave just like light. If a radio signal was sent from New York to California, it would take approximately 0.016 seconds to complete the journey. Multiplying velocity by time, 186,000 miles per second by 0.016 seconds, gives us a distance of 3000 miles; the distance between New York and California (Fig. 6-1). In other words, a ruler which consists of a radio signal and a clock (to read .016 seconds), gives us the same reading for distance that we would expect to get if we could stretch a measuring tape made of cloth, wood, or steel from New York to

California. We will return to this rather obvious point in the discussion which follows.

Imagine looking up and watching our friends in their Sun-Earth System which, for the moment, is standing perfectly still relative to us (Fig. 6-2). A distance of 93,000,000 miles separates their Sun from their Earth. If our friend on the Sun presses a switch and turns on the Sun at 6:00, a photon of light will take 8 minutes to travel to their Earth.

If our friends' Sun-Earth System travels across our line of vision at a speed of perhaps 100,000 miles per second, the photon, as we already know, will fail to reach the Earth in 8 minutes. While the photon travels towards the Earth, the Earth moves away from the photon. Imagine now that our friends remedy the situation by moving their Earth closer to their Sun and call their new shortened distance 93,000,000 miles. Having adjusted the distance between them, our friends, who are still traveling at 100,000 miles per second, can now repeat their experiment successfully (Fig. 6-3). At exactly 6 o'clock they turn on the Sun, and at 6:08 the man on the Earth looks up and sees a photon of light; a

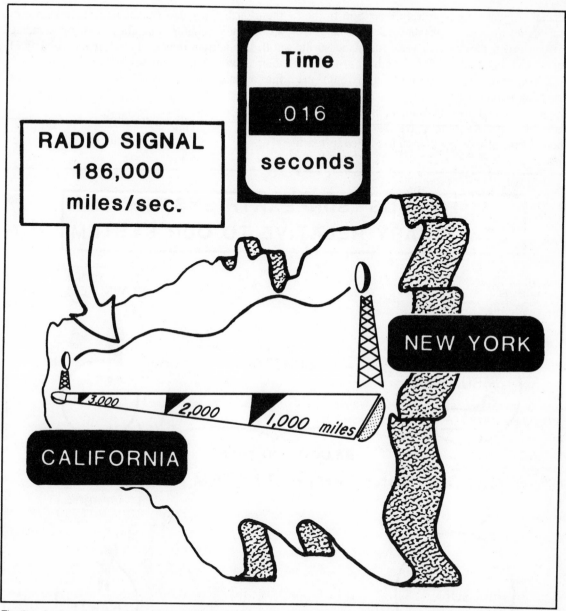

Fig. 6-1. A ruler made of some material like cloth or wood provides the same measurement as a ruler consisting of a radio signal and a clock.

photon which he claims has traveled 93,000,000 miles, but which we will claim has traveled some distance shorter than 93,000,000 miles. If our friend on the Sun sent a radio signal at 6:00, then his colleague would receive that signal at 6:08. Our friends will no doubt claim that the radio signal, like the photon, traveled 93,000,000 miles. We, however, will claim that it traveled some distance shorter than 93,000,000 miles. As we said earlier, a ruler consisting of a radio signal and a clock must provide the same measurement as a ruler made of cloth, wood, or steel. Consequently, when they stretch a measuring tape from their Sun to their Earth, they must get 93,000,000 miles. As far as we are concerned, they must therefore use a measuring tape which has contracted in length.

To review the situation, our friends are traveling at a constant velocity of 100,000 miles per second across our line of vision. At 6:00 they turn on their Sun, and a photon starts traveling towards their Earth. At 6:08 the photon has failed to reach their Earth. Since they cannot detect their motion in space, they maintain that they must be standing still. In addition, they cannot alter the velocity of their photon. Since they are well acquainted with the formula

$$\text{Time} = \frac{\text{Distance}}{\text{Velocity}}$$

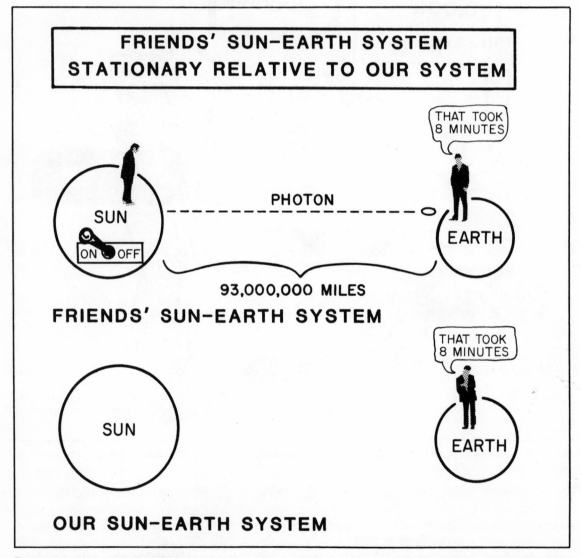

Fig. 6-2. If our friends' Sun-Earth System is stationary relative to our System, then we will find that a photon takes eight minutes to travel from their Sun to their Earth.

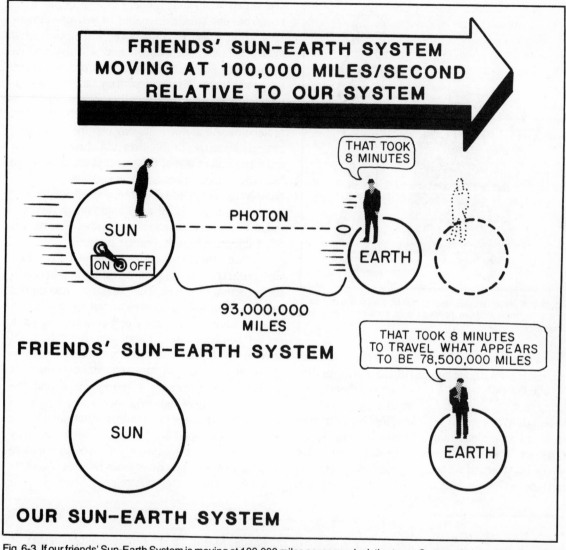

Fig. 6-3. If our friends' Sun-Earth System is moving at 100,000 miles per second relative to our System, then our friends will find it necessary to move their Earth closer to their Sun in order for the photon to take eight minutes to travel from their Sun to their Earth. Our friends will continue to maintain that they are 93,000,000 miles apart, however, we will maintain that they are only 78,500,000 miles apart.

and they know that they cannot tamper with the velocity of the photon, they decide to disregard time, and shorten the distance between them. In doing so, however, they are forced to call their new shortened distance 93,000,000 miles. This allows them to use their formula to get the 8 minutes required for a photon to travel from their Sun to their Earth.

$$\text{Time} = \frac{\text{Distance}}{\text{Velocity}} = \frac{93{,}000{,}000 \text{ miles}}{186{,}000 \text{ miles/sec.} \times 60 \text{ sec./min}}$$

$$\text{Time} = \text{approximately 8 minutes}$$

Fig. 6-4. A shrunken ruler pulled out of a washing machine differs significantly from one which is smaller in only one dimension.

one which is smaller in every dimension, is different from one which is smaller in only one dimension (Fig. 6-4). Furthermore, the notion of a measuring tape contracting in one of its dimensions is far more subtle than the notion of a shrunken measuring tape. When a measuring tape shrinks, its molecules come closer together; when we say it contracts in one dimension, we imply that the molecules and atoms that compose the measuring tape also contract in the direction of motion. If we were to think of one of the atoms in the measuring tape as a small sphere, we would find that upon contraction it would resemble a small ellipsoid— something like a small egg. Maintaining that distance undergoes contraction, implies that matter itself undergoes contraction.

Notice, that in our entire discussion of contraction, nothing happens in a direction perpendicular to the direction of travel. The distance between our friends' Earth and Sun contracts, but no contraction takes place in any direction perpendicular to a line drawn between their Earth and their Sun. Of course, all this sounds vaguely familiar from our discussion of the equations written by Lorentz. It should therefore come as no surprise that the shortened distance between our friends' Earth and Sun can be computed by a familiar equation; the first equation developed by Lorentz. The reader may recall that this first eqution was used to compute the shortened distance between the beam splitter and the mirror located upstream.

From our point of view, what they now call 93,000,000 miles, we see as a distance shorter than 93,000,000 miles. Furthermore, we see them measuring that distance with a ruler that has contracted in length. When we say that the ruler has contracted in length, we mean that it is shorter only in the direction of motion. This is unlike the situation we would encounter if we were to pull a cloth measuring tape out of a washing machine. A shrunken ruler,

$$
\begin{pmatrix} \text{New (shortened)} \\ \text{distance between} \\ \text{beam splitter} \\ \text{and upstream} \\ \text{mirror} \end{pmatrix} = \sqrt{\dfrac{\left(\begin{array}{l}\text{Velocity of Michelson-Morley} \\ \text{apparatus relative to ether}\end{array}\right)^2}{(\text{Velocity of light})^2}} \begin{pmatrix} \text{Measured distance} \\ \text{between beam} \\ \text{splitter and} \\ \text{upstream mirror} \end{pmatrix}
$$

Rewriting this equation in terms of our two Sun-Earth Systems, we obtain:

$$
\begin{pmatrix} \text{Shortened distance} \\ \text{between our friends'} \\ \text{Earth and Sun} \end{pmatrix} = \sqrt{1 - \dfrac{\left(\begin{array}{l}\text{Velocity of our friends'} \\ \text{Sun-Earth System relative} \\ \text{to us}\end{array}\right)^2}{(\text{Velocity of light})^2}} \begin{pmatrix} \text{Distance between} \\ \text{our friends' Earth} \\ \text{and Sun when they} \\ \text{are stationary} \\ \text{relative to us} \end{pmatrix}
$$

Substituting numbers into this equation gives us:

$$\text{Shortened distance between our friends' Earth and Sun} = \sqrt{1 - \frac{(100,000)^2}{(186,000)^2}}\ (93,000,000)$$

$$= 78,500,000 \text{ miles}$$

The shortened distance between our friends' Earth and Sun equals approximately 78,500,000 miles.

In other words, when our friends are traveling across our line of vision at 100,000 miles per second, the shortened distance between their Earth and their Sun will appear to us to be 78,500,000 miles. Furthermore, if they were to stretch a ruler from their Sun to their Earth, we would notice that one end of the ruler reads "zero" and the other end reads "93,000,000"; however, we would also see that the distance between the "zero" and "93,000,000" is 78,500,000 miles. Obviously, we would be

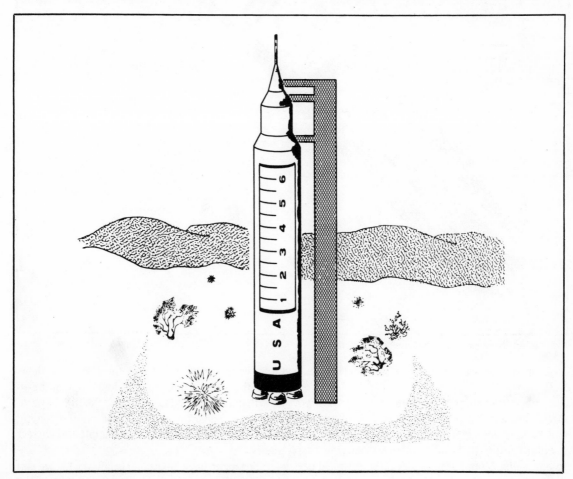

Fig. 6-5. Contraction takes place only in the direction of motion.

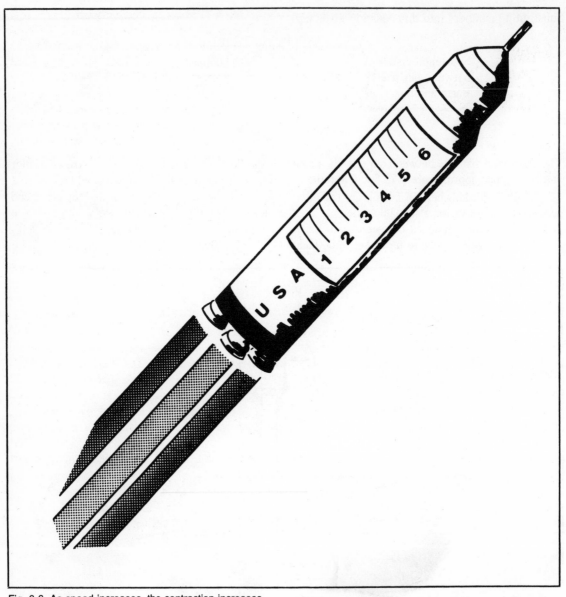

Fig. 6-6. As speed increases, the contraction increases.

looking at a contracted ruler (Figs. 6-5, 6-6, and 6-7).

Since no contraction takes place in the two directions perpendicular to the direction of motion, the second and third equations developed by Lorentz remain the same. In other words, referring to our friends' Sun-Earth System, the distance between any two points in a direction perpendicular to

their direction of travel will be the same regardless of whether they are moving or not. We find, therefore, that even though Lorentz was the first to write the correct equations for the contraction of length, Albert Einstein was the first to supply the correct theory.

So far, we have described in detail what we will see if our friends cross our line of vision at a con-

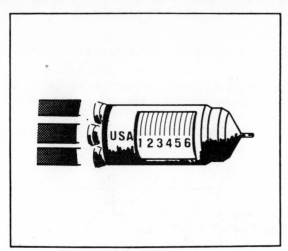

Fig. 6-7. Near the speed of light the contraction becomes profound.

look at them, and they find the same when looking at us, each of us will find that everything appears to be normal within our own Sun-Earth System. When they stretch a measuring tape between their Earth and their Sun, it will appear to them to be 93,000,000 miles long even though it appears to be 78,500,000 miles long to us. They will not complain that things appear to be contracted in the direction of travel, because, as far as they are concerned, they cannot even prove that they are moving. The same holds true when we stretch a measuring tape between our Earth and our Sun. We, too, will not complain of any contraction because we are also unable to detect our motion in space. To each of us, contraction in the direction of motion will be apparent only in the other Sun-Earth System.

Let us return once again to our friends who are still traveling at 100,000 miles per second across our line of vision. From our vantage point, as we already know, the distance between them appears to be 78,500,000 miles. Suppose now that they begin traveling across our line of vision at 150,000 miles per second. At exactly 6:00 our friend on the Sun presses the switch and a photon of light starts traveling to his colleague on Earth. At 6:08, as you might expect, the photon has still not reached the Earth. The Earth, of course, has been moving away from the photon during the eight minute interval. Again, the situation is remedied by moving the Earth closer to the Sun (Fig. 6-8). The perceived distance between our friends will become smaller than 78,500,000 miles; a distance which can be

stant velocity. But what will they see when they look at us? The strange thing about all this is that they will see everything we saw! When they look at us, they will maintain that they are standing still, and that we are crossing their line of vision at 100,000 miles per second. If we were to turn on our Sun at 6:00 they too will find that a photon fails to reach our Earth at 6:08. We, of course, will react the way they did, and shorten the distance between our Earth and our Sun. They will end up seeing a shorter version of 93,000,000 miles just the way we did when we looked at them.

Even though we find that distance and matter are contracted in the direction of motion when we

$$\text{Shortened distance between our friends' Earth and Sun} = \sqrt{1 - \frac{\left(\begin{array}{c}\text{Velocity of our friends'}\\ \text{Sun-Earth System relative}\\ \text{to us}\end{array}\right)^2}{(\text{Velocity of light})^2}} \left(\begin{array}{c}\text{Distance between our}\\ \text{friends' Earth and}\\ \text{Sun when they are}\\ \text{stationary relative}\\ \text{to us}\end{array}\right)$$

$$\text{Shortened distance between our friends' Earth and Sun} = \sqrt{1 - \frac{(150,000)^2}{(186,000)^2}}\ (93,000,000)$$

$$\text{Shortened distance between our friends' Earth and Sun} = \text{approximately } 55,000,000 \text{ miles}$$

determined by applying our formula. Now, when we look at our friends, we see a distance of 55,000,000 miles between their Earth and their Sun. When they look at us, they find that the same distance separates our Earth from our Sun.

Let us go to the ultimate extreme and imagine that our friends are crossing our line of vision at the speed of light itself—186,000 miles per second.

Once again, another try of their experiment in which they turn on the Sun at 6:00 will lead to a shortening of the distance between their Earth and their Sun. But this time it will be necessary to move the Earth right up next to the Sun, thereby shortening the distance between them to zero (Fig. 6-9). This also becomes evident when we use our formula.

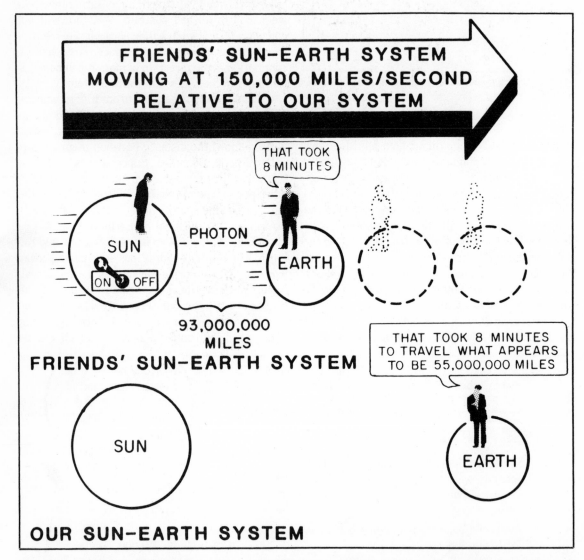

Fig. 6-8. If our friends' Sun-Earth System is moving at 150,000 miles per second relative to our System, then, once again, it will be necessary for our friends to move their Earth closer to their Sun in order for the photon to take eight minutes to travel from their Sun to their Earth. Our friends will still maintain that they are 93,000,000 miles apart, but we will claim that they are now only 55,000,000 miles apart.

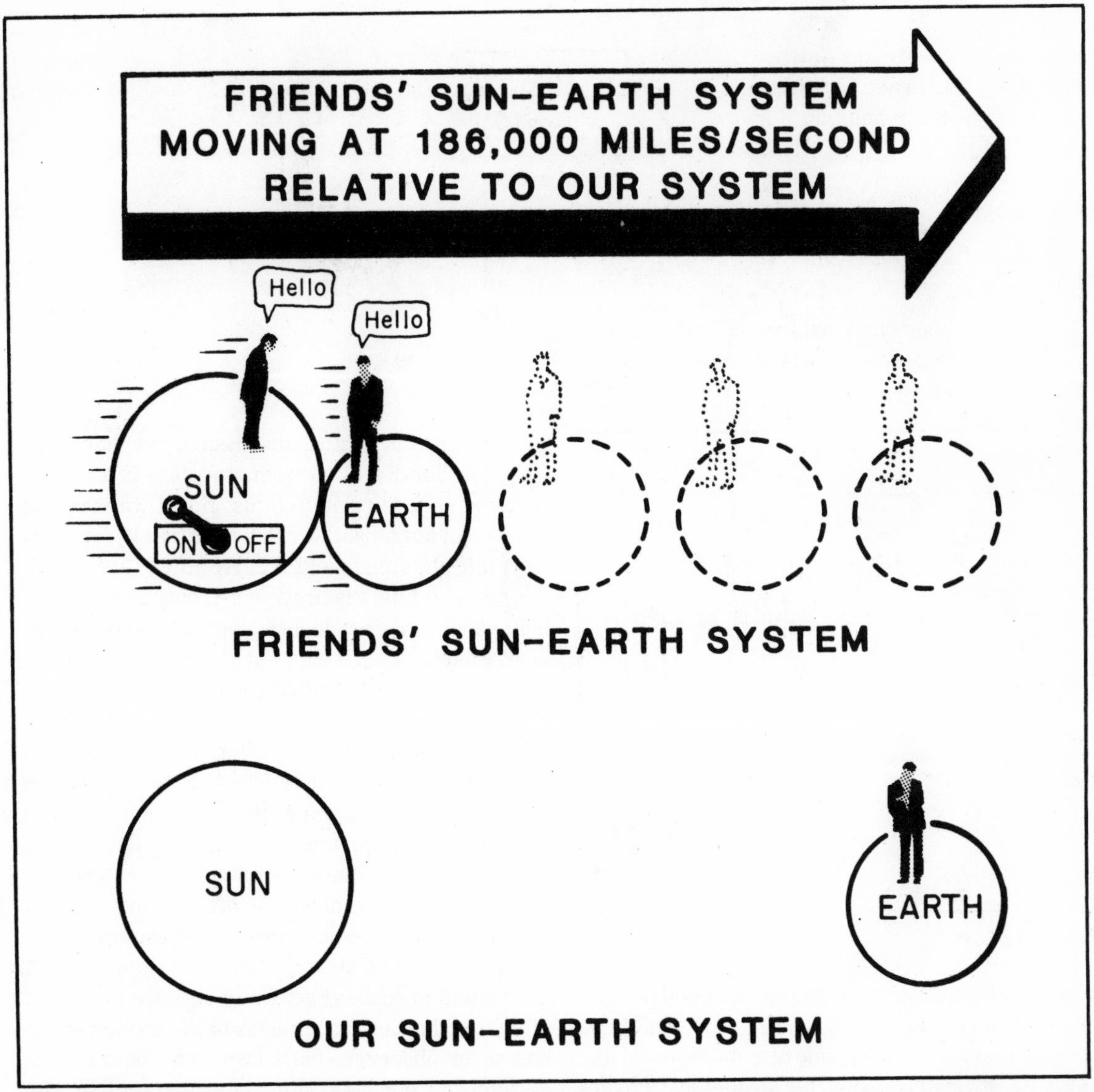

**FRIENDS' SUN-EARTH SYSTEM MOVING AT 186,000 MILES/SECOND RELATIVE TO OUR SYSTEM**

Hello

Hello

SUN

ON OFF

EARTH

**FRIENDS' SUN-EARTH SYSTEM**

SUN

EARTH

**OUR SUN-EARTH SYSTEM**

Fig. 6-9. If our friends' Sun-Earth System is moving at 186,000 miles per second relative to our System, our friends will find that they have to move their Earth next to their Sun if our Earthbound friend is to have any chance at all of seeing the photon. Since they cannot detect their motion in space, however, they have no reason to believe that they are moving. They will therefore continue to believe that they are 93,000,000 miles apart, while we maintain that they are close enough to say "Hello" to each other.

$$\begin{pmatrix}\text{Shortened} \\ \text{distance between} \\ \text{our friends' Earth} \\ \text{and Sun}\end{pmatrix} = \sqrt{1 - \dfrac{\left(\begin{array}{c}\text{Velocity of our friends'} \\ \text{Sun-Earth System relative} \\ \text{to us}\end{array}\right)^2}{(\text{Velocity of light})^2}} \left(\begin{array}{c}\text{Distance between our} \\ \text{friends' Earth and} \\ \text{Sun when they are} \\ \text{stationary relative} \\ \text{to us}\end{array}\right)$$

$$\text{Shortened distance between our friends' Earth and Sun} = \sqrt{1 - \frac{(186,000)^2}{(186,000)^2}} \quad (93,000,000)$$

$$\text{Shortened distance between our friends' Earth and Sun} = \sqrt{1 - 1} \quad (93,000,000)$$

$$\text{Shortened distance between our friends' Earth and Sun} = 0 \text{ miles}$$

Fig. 6-10.

When our friends are traveling at the speed of light, length shrinks to zero. A measuring tape would also shrink to zero. The tape, made of some material like cloth, wood, or steel, would get smaller and smaller as it traveled faster and faster relative to us; finally, at the speed of light, it would disappear altogether.

During our entire discussion, we have imagined a Sun-Earth System traveling at ever increasing speeds relative to us. But if we stand on the side of a road and watch a car drive by, it too is traveling at some speed relative to us (Fig. 6-10). Of course, it might be traveling at only 55 miles per hour which is a great deal slower than 100,000 miles per second. Nevertheless, it is traveling at some speed relative to us, and consequently we can apply our formula to calculate the new, smaller length of the car. We would find, however, that the change in length was so small that it would certainly never be visible to the human eye. In order for us to even hope to see an object get smaller in its direction of travel, we would have to watch it traveling at a speed approaching thousands of miles per second.

What we have described so far is strange indeed! It all comes about because we cannot detect our motion in space (Fact One); because the speed of light is the same regardless of the motion of the source or observer (Fact Two); and because we decide to accept the results, however strange they may be, which emerge when we utilize these two facts in a logical sequence of events.

**Chapter 7**

# Time

L ET US RETURN ONCE AGAIN TO OUR FRIENDS ON
their Sun-Earth System. Again, we imagine
that they are standing perfectly still relative to us,
and that a distance of 93,000,000 miles separates
the two of them (Fig. 7-1). At 6:00, our friend on the
Sun presses a switch and a photon of light starts
traveling towards the Earth. At 6:08 our friend on
the Earth looks up and sees the photon of light. We
can now envision our friends on their Sun-Earth
System traveling across our line of vision at a speed
of 100,000 miles per second. At 6:00, the man on the
Sun presses a switch and a photon starts traveling
towards the Earth because during the time the
photon was traveling the Earth was moving away
from the photon. The situation by now is a familiar
one. Our friends cannot prove that they are moving
through space, and they cannot alter the speed of
their photon. Yet, somehow, circumstances must be
changed so that the photon takes 8 minutes to travel
to their Earth. Our friends, once again, examine the
formua:

$$\text{Time} = \frac{\text{Distance}}{\text{Velocity}}$$

and, being unable to change the velocity of the
photon, decide in this instance to ignore distance
and change time. In other words, they leave the
Earth where it is and take their clocks to the jeweler
to have them slowed down. If they now repeat their
experiment while traveling across our line of vision
at 100,000 miles per second, they will find that they
have solved their problem (Fig. 7-2). At 6:00 they
turn on their Sun, and at 6:08, according to their
newly adjusted clocks, the photon reaches their
Earth. During this second attempt, we will witness
the photon taking eight lazy, slow minutes to travel
from their Sun to their Earth. As a matter of fact,
Einstein reasoned that if we used the fourth equation written by Lorentz, the one Lorentz had so
much trouble trying to explain, we could compute
the amount of time it takes for the photon to travel
from the Sun to the Earth according to our clocks.
But first, let us recall that troublesome equation:

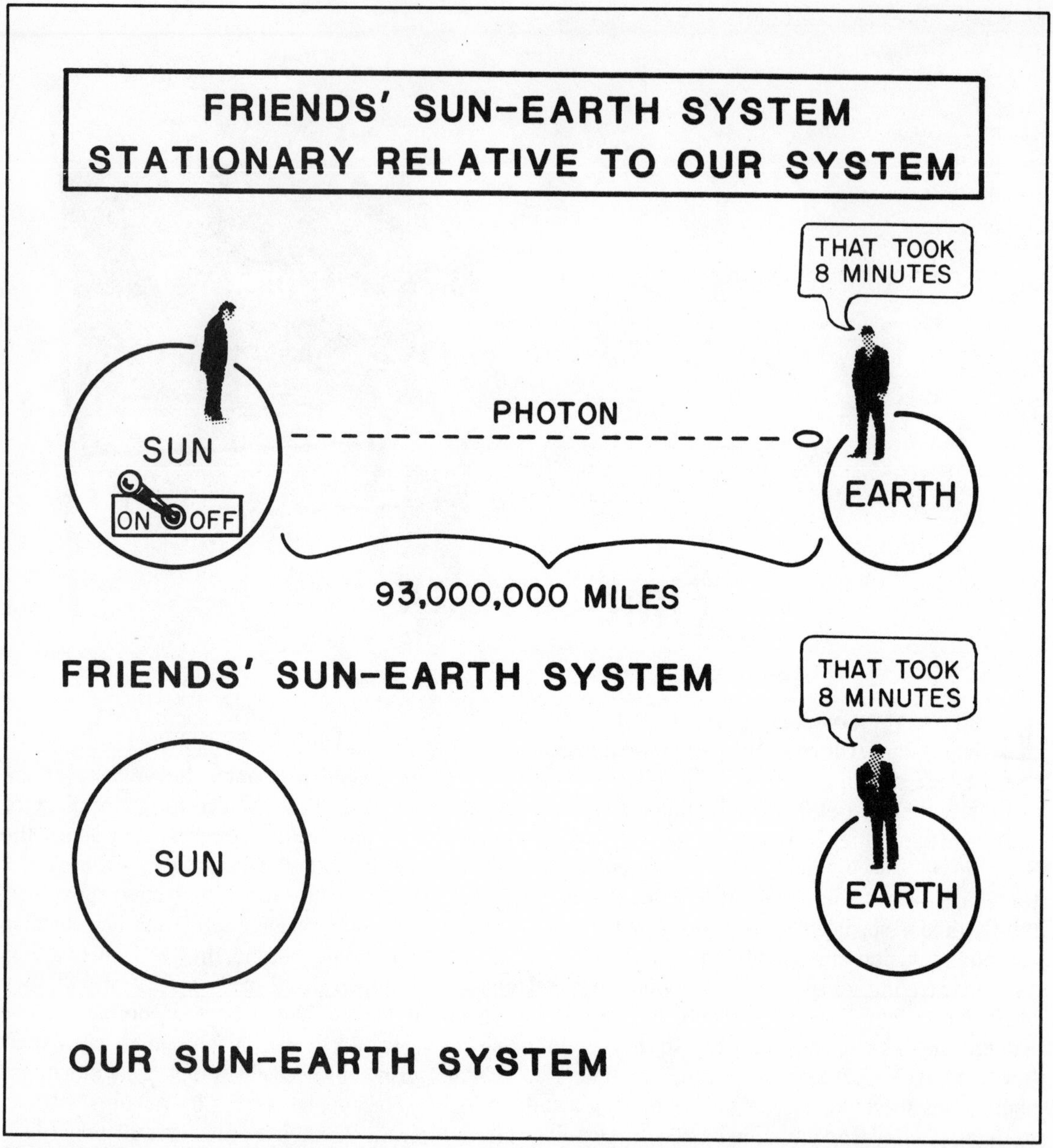

Fig. 7-1. If our friends' Sun-Earth System is stationary relative to our System, then we will find that a photon takes eight minutes to travel from their Sun to their Earth.

$$\begin{pmatrix}\text{``Artificial'' time for}\\\text{light beam to complete}\\\text{round trip journey}\\\text{between beam splitter}\\\text{and upstream mirror}\end{pmatrix} = \sqrt{1 - \dfrac{\begin{pmatrix}\text{Velocity of Michelson-}\\\text{Morley apparatus}\\\text{relative to ether}\end{pmatrix}^2}{(\text{Velocity of light})^2}} \begin{pmatrix}\text{Measured time for light}\\\text{beam to complete round}\\\text{trip journey between}\\\text{beam splitter and}\\\text{upstream mirror}\end{pmatrix}$$

Rewriting the equation for our two Sun-Earth Systems, we obtain:

Time for photon to travel from our friends' Sun to our friends' Earth according to their clocks $= \sqrt{1 - \dfrac{\left(\begin{array}{l}\text{Velocity of our friends'}\\ \text{Sun-Earth System}\\ \text{relative to us}\end{array}\right)^2}{(\text{Velocity of light})^2}}\left(\begin{array}{l}\text{Time for photon to}\\ \text{travel from our friends'}\\ \text{Sun to our friends'}\\ \text{Earth according}\\ \text{to our clocks}\end{array}\right)$

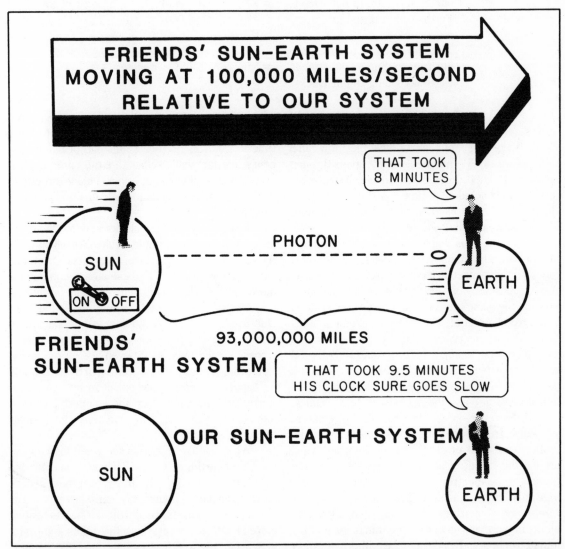

Fig. 7-2. If our friends' Sun-Earth System is moving at 100,000 miles per second relative to our System, then our friends will find it necessary to have their clocks slowed down. They will now maintain that the photon takes eight minutes to travel from their Sun to their Earth according to their newly adjusted clocks. However, we will maintain that the photon took nine and a half minutes according to our clocks.

Substituting the proper numbers into this equation gives us:

$$8 = \sqrt{1 - \frac{(100{,}000)^2}{(186{,}000)^2}} \quad \left( \begin{array}{l} \text{Time for photon to travel from our} \\ \text{friends' Sun to our friends'} \\ \text{Earth according to our clocks} \end{array} \right)$$

$$\begin{array}{l} \text{Time for photon to travel from} \\ \text{our friends' Sun to our friends'} \\ \text{Earth according to our clocks} \end{array} = \sqrt{\dfrac{8}{1 - \dfrac{(100{,}000)^2}{(186{,}000)^2}}}$$

$$\begin{array}{l} \text{Time for photon to travel from} \\ \text{our friends' Sun to our friends'} \\ \text{Earth according to our clocks} \end{array} = \text{approximately } 9\tfrac{1}{2} \text{ minutes}$$

It is no wonder that we see the photon taking eight lazy, slow minutes to travel from their Sun to their Earth. According to our clocks, what they now call 8 minutes we call 9½ minutes. Their travels through space at 100,000 miles per second relative to us leads us to conclude that time has slowed down for them—the minute hand on their clock turns more slowly than the minute hand on ours.

In the previous chapter, we made the point that all kinds of rulers must provide the same measurement if they are used to measure the same distance. A ruler consisting of a clock and a radio signal must provide the same measurement as a ruler made of cloth, wood, or steel. The same requirement holds true for different types of clocks. A watch with a sweep second hand, a battery operated digital watch, or a pendulum clock must all record the passage of time equally if we are to have any confidence at all in using these devices. A mechanical device, however, is not the only kind of clock in existence. In fact, anything that undergoes periodic motion is a type of clock. Since electrons whirl around the nuclei of atoms at a regular rate, the motion of the electrons constitute an unusual, but legitimate clock. Since your heart beats at a regular rate, it too constitutes an unusual, but legitimate clock. In our everyday lives, the things we do take minutes or hours; they do not take millions of whirls of an electron or hundreds of heart beats. Nevertheless, if an electron always circles around a nucleus the same number of times each second, or if

a heart beats the same number of times each minute, there is no reason why we could not record time in "whirls" or "beats" instead of minutes or hours. The fact that all clocks are the same, and that there are so many unusual clocks, leads directly to the strangest conclusions of all in the Theory of Relativity.

Having our friends' clocks slowed down implies far more than a simple mechanical adjustment. A clock, like everything else, is composed of atoms; and even the electrons in these atoms would move more slowly. The movement of electrons in the batteries of a digital clock would also be slower. Our friends would be moving and talking more slowly, their hearts would beat more slowly, and as a consequence they would even age more slowly. Everything in our friends' Sun-Earth System would be happening more slowly than in our own Sun-Earth System. Since even the smallest particles, like electrons, would be moving more slowly, matter itself is affected.

What would our friends see when they look at us? The same things we see! They will maintain that everything in our Sun-Earth System moves more slowly than it does in their System. They will see us moving and talking more slowly, our hearts beating more slowly, and everything that takes eight minutes according to our clocks will take 9½ minutes according to their clocks. Now we are the ones who visited the jeweler to have our clocks slowed down. Of course, while they assert that everything is moving at a normal speed in their System, we assert that the same is true in our System.

As strange as everything appears to be when each of us looks at the other, the real complications begin when we start traveling from one system to the other. Remember, when we look at our friends, we maintain that we are standing still while they are traveling at 100,000 miles per second relative to us. In other words, they are traveling 100,000 miles per second faster than we are. If were to visit them, we would have to get into a rocketship and attain a velocity of 100,000 miles per second before we

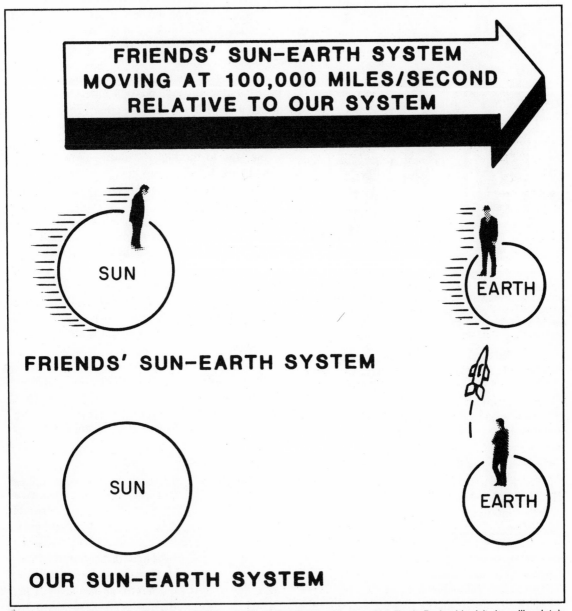

Fig. 7-3. Twin paradox. Jon travels to our friends' Sun-Earth System and lands on their Earth. During his visit, Jon will maintain that his twin brother Ron is aging more slowly, and Ron will maintain that Jon is aging more slowly.

Fig. 7-4. When Jon returns home, he will greet his twin brother Ron who is now his older brother. During the journey, Jon aged more slowly because he is the brother who accelerated to the higher velocity.

This leads to the story of the twin brothers, Jon and Ron, who live with us on our Sun-Earth System. One day Jon gets into a rocketship and blasts off (Fig. 7-3). He accelerates up to a velocity of 100,000 miles per second, travels to our friends' Sun-Earth System, and lands on their Earth. During his stay on our friends' Earth, Jon travels 100,000 miles per second faster than he did here on our Earth. When Ron looks at his brother, Jon appears to be aging more slowly. But when Jon looks back at his brother, Ron appears to be the one who is aging more slowly.* This phenomenon is known as the Twin Paradox. How is it possible for each brother to be aging more slowly than the other? The answer is of course that it is not possible! Jon is the brother who is aging more slowly because he is the one who accelerated to the higher velocity. When Jon returns home, he will greet his "twin" brother who during the journey became his older brother (Fig. 7-4).

Let us now suppose that our friends' Sun-Earth System speeds up to 150,000 miles per second (Fig. 7-5). If the man on the Sun presses the switch at 6:00, the photon will once again fail to reach the Earth by 6:08. Again, our friends must have their clocks slowed down by the jeweler. After having their clocks adjusted, the man on the Sun can press the switch at 6:00, and the man on the Earth will see the photon at 6:08.

With the increase in velocity, we will undoubtedly see things moving even more slowly than we did before. The Lorentz equation tells us the new, longer time for the photon to travel from our friends' Sun to our friends Earth according to our clocks:

---

*Our explanation of what Jon and Ron see is limited to that predicted by the Theory of Relativity. A complete explanation of what they would actually see requires a fairly lengthy discussion of a number of phenomena. These are not discussed here because they do not pertain to the Theory of Relativity.

could land on their Earth. Our friends, of course, face the same situation. In looking at us, they maintain that they are standing still and that we are traveling 100,000 miles per second faster than they are. If they want to visit us, they too will have to get into a rocketship and attain a velocity of 100,000 miles per second before they can land on our Earth. In either case, it is necessary to get into a rocketship and accelerate from a velocity of zero to a velocity of 100,000 miles per second before visiting the other Sun-Earth System.

$$
\begin{pmatrix} \text{Time for photon to} \\ \text{travel from our friends'} \\ \text{Sun to our friends' Earth} \\ \text{according to their clocks} \end{pmatrix} = \sqrt{1 - \frac{\begin{pmatrix} \text{Velocity of our friends'} \\ \text{Sun-Earth System} \\ \text{relative to us} \end{pmatrix}^2}{(\text{Velocity of light})^2}} \begin{pmatrix} \text{Time for photon to} \\ \text{travel from our friends'} \\ \text{Sun to our friends'} \\ \text{Earth according} \\ \text{to our clocks} \end{pmatrix}
$$

$$8 = \sqrt{1 - \frac{(150{,}000)^2}{(186{,}000)^2}} \left( \begin{array}{l} \text{Time for photon to travel from our} \\ \text{friends' Sun to our friends' Earth} \\ \text{according to our clocks} \end{array} \right)$$

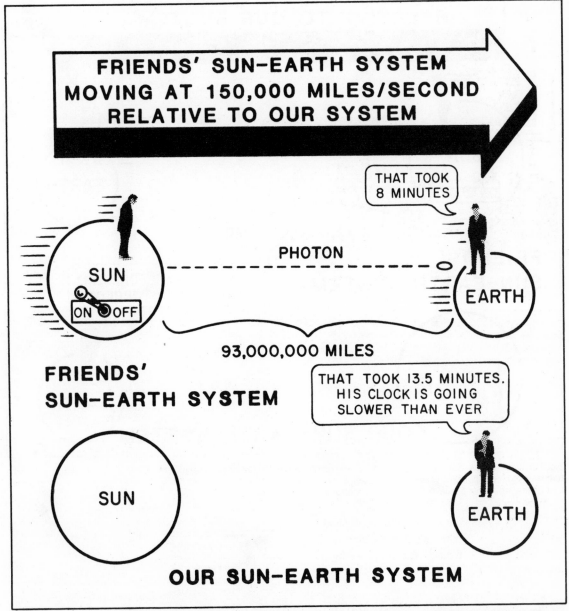

Fig. 7-5. If our friends' Sun-Earth System is moving at 150,000 miles per second relative to our System, then, once again, our friends will find it necessary to have their clocks slowed down. Our friends will of course maintain that the photon took eight minutes to travel from their Sun to their Earth according to their clocks. However, we will maintain that the photon took thirteen and a half minutes according to our clocks.

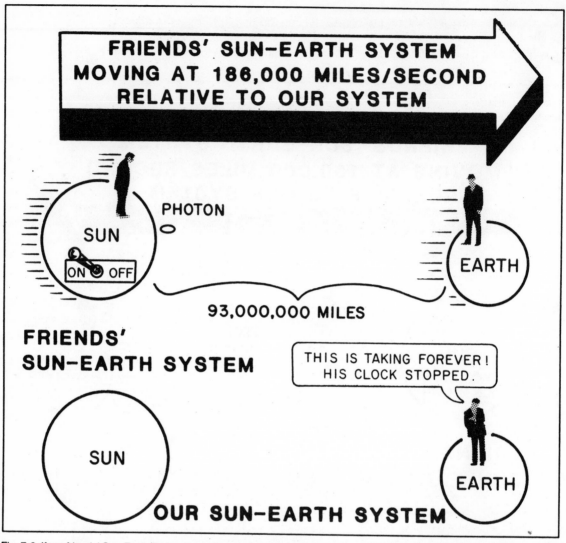

Fig. 7-6. If our friends' Sun-Earth System is moving at 186,000 miles per second relative to our System, our friends will find it necessary to have their clocks stopped completely. But since the photon never leaves the surface of their Sun, we will maintain that according to our clocks the travels of the photon are taking forever.

$$\text{Time for photon to travel from our friends' Sun to our friends' Earth according to our clocks} = \frac{8}{\sqrt{1 - \frac{(150,000)^2}{(186,000)^2}}}$$

Time for photon to travel from our friends' Sun to our friends' Earth according to our clocks = Approximately 13½ minutes

Of course, when our friends look at us, we will be the ones who appear to be moving even more slowly. Eight minutes on our clocks will correspond to 13½ minutes on their clocks.

We now go to the ultimate extreme, and imagine that our friends are passing us on their Sun-Earth System at the speed of light itself—186,000 miles per second (Fig. 7-6). At 6:00, the man on the Sun presses the switch, and this time the photon never even leaves the Sun. Our friends must now take their clocks to the jeweler and have them stopped completely. At the speed of light, time stands still.

As our friends moved faster and faster through space, the passage of eight minutes for them corresponded to the passage of first 9½ minutes, then 13½ minutes, and now finally an infinite amount of time. It appears now that the photon will take forever to travel from their Sun to their Earth. Indeed, the Lorentz equation confirms this.

sheer speculation, because according to the Theory of Relativity we can never travel at the speed of light; but more about this later.

If time appears to move more slowly on a Sun-Earth System which is traveling at some high speed relative to us, then it moves more slowly for anybody or anything that is moving relative to us. As we stand outside, everybody who is driving by or flying by is aging more slowly than we are. Their speed, however, is so small relative to the speed of light that the difference between how quickly they age and how quickly we age is negligible. They would all have to be traveling at some speed close to the speed of light before we would notice that they are aging at a substantially slower rate than we are. Everybody, in a very literal sense, lives according to their very own private time.

We close this chapter, as we did the last, with the reminder that all of these strange events occur because we cannot detect our motion in space (Fact

$$\begin{pmatrix} \text{Time for photon to} \\ \text{travel from our friends'} \\ \text{Sun to our friends' Earth} \\ \text{according to their clocks} \end{pmatrix} = \sqrt{1 - \frac{\begin{pmatrix} \text{Velocity of our friends'} \\ \text{Sun-Earth System} \\ \text{relative to us} \end{pmatrix}^2}{(\text{Velocity of light})^2}} \begin{pmatrix} \text{Time for photon to} \\ \text{travel from our friends'} \\ \text{Sun to our friend's} \\ \text{Earth according} \\ \text{to our clocks} \end{pmatrix}$$

$$8 = \sqrt{1 - \frac{(186,000)^2}{(186,000)^2}} \begin{pmatrix} \text{Time for photon to travel from our} \\ \text{friend's Sun to our friends' Earth} \\ \text{according to our clocks} \end{pmatrix}$$

$$\begin{pmatrix} \text{Time for photon to travel} \\ \text{from our friends' Sun to our} \\ \text{friends' Earth according to} \\ \text{our clocks} \end{pmatrix} = \frac{8}{\sqrt{1 - \frac{(186,000)^2}{(186,000)^2}}}$$

$$\begin{pmatrix} \text{Time for photon to travel} \\ \text{from our friends' Sun to our} \\ \text{friends' Earth according to} \\ \text{our clocks} \end{pmatrix} = \text{infinity}$$

If time stands still for our friends, their hearts must stop beating. Of course, in ordinary circumstances they would die. In this situation, however, that does not happen because their hearts have not stopped beating for a period of time. All of this, however, must be relegated to the realm of

One); because the speed of light is the same regardless of the motion of the source or observer (Fact Two); and because we decide to accept whatever strange results emerge when utilizing these two facts in a logical sequence of events.

# The
# Time Machine

OVER THE YEARS, SCIENCE FICTION WRITERS have had a heyday writing about time. The Theory of Relativity certainly added fuel to the fire because the idea of traveling back in time could easily be implied from the logic of the theory itself. If clocks slow down as they approach the speed of light, and clocks stop at the speed of light, then clocks must go backwards once they exceed the speed of light. In other words, if we could go faster than the speed of light, we would go back in time. This intriguing idea is not true for reasons which we will discuss shortly. In the meantime, let us at least explore some of the implications in this line of reasoning.

Suppose we wanted to visit Isaac Newton. In order to appreciate the visit, it is essential that we remain as old as we are right now. One way to do this might be to travel to the Moon. While waiting patiently on the Moon, we pack the Earth away in a giant rocket and send it off on a journey through space traveling faster than the speed of light (Figs. 8-1 and 8-2). Time on the Earth will go backwards,

previous generations will come to life, and eventually Isaac Newton will once again walk the Earth. We stop the rocket, unpack the Earth, and visit with Isaac Newton. When we are ready to return to our own time, we will have to get into a rocket and travel about the universe at some speed close to, but not exceeding the speed of light (Figs. 8-3 and 8-4). Under these conditions, we will age more slowly while people on Earth age more quickly. Generations of people on Earth will pass on until finally our own generation is reborn. We can then leave our rocket and return home to greet our contemporaries.

Or, consider for a moment the day the Sun blows up! Eight minutes after the actual event, we see this stupendous explosion in the sky. We are so fascinated that we decide to see it all over again. We jump into a rocket and head for Mars at a speed faster than light (Fig. 8-5). We overtake the light of the explosion and land on Mars. Once out of our rocket, we have only to look up in the sky and see the Sun blow up all over again.

Fig. 8-1. To visit Isaac Newton we must load the Earth in a spaceship.

Fig. 8-3. When the spaceship returns we can unload the Earth and visit with Isaac Newton.

Such are the flights of imagination in the world of relativity; or to be more accurate, outside of the world of relativity. The speed chart, Fig. 8-6, summarizes our movement in time at various speeds. The cross-hatched areas highlight those speeds, which according to current scientific thinking, must forever remain within the realm of science fiction. Right now we are moving into the future at a rate of

24 hours per day. The chart tells us that if we could move through the universe at some faster rate, we would move into the future more slowly. If we move at the speed of light, we stand still in time; and, if we move faster than the speed of light, we move back in time. At the other extreme, the chart shows that if we could travel through the universe more slowly than we are now, presumably by stopping the world

Fig. 8-2. The spaceship must travel faster than the speed of light.

Fig. 8-4. In order to return to our own time, we must board the spaceship and travel at some speed slightly slower than the speed of light.

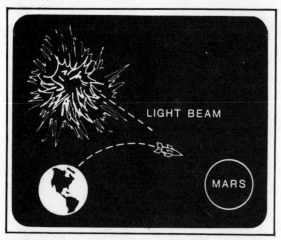

Fig. 8-5. Having seen the Sun explode once, it will be necessary to travel faster than light in order to see it explode all over again.

and getting off, we could move into the future more quickly than the present rate of 24 hours per day. Either extreme, as indicated by the cross-hatched area, is impossible.

Let us see what happens when we try to step into the future, or, in other words, move into the future at a rate faster than 24 hours per day. We begin by finding a place in the universe which is traveling more slowly than the Earth. Mars is a good candidate because it is traveling around the Sun more slowly than the Earth. The Earth travels around the Sun at a speed of 18.5 miles per second, whereas Mars travels at 15 miles per second. Before we can even begin to contemplate our journey, we must contend with Fact One in the Theory of Relativity. Our inability to detect our motion in space forces us to conclude that for all practical purposes we are standing still and Mars is moving

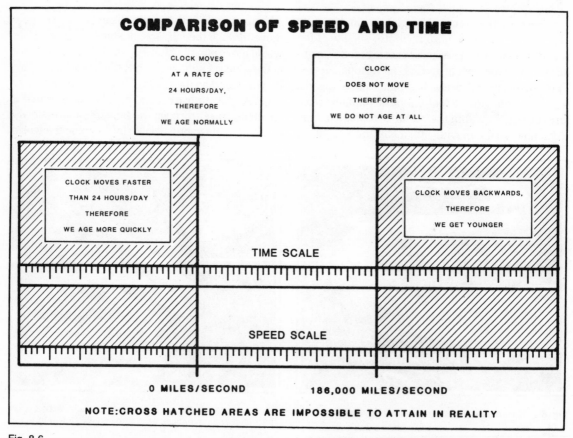

Fig. 8-6.

relative to us. Consequently, we must get into a rocket and accelerate to some high velocity in order to get to slow moving Mars. We cannot stop the world, get off, and watch it hurtle out of sight, while we wait around for the arrival of Mars; nor can we get into a rocket and decelerate our way to Mars. From here on Earth, everything is moving faster than we are because of Fact One. It is simply impossible to go any slower than we are already going, and we can therefore never even hope to move into the future more quickly.

Now, let us look at the other extreme and see what happens when we try to travel faster than the speed of light. We already know that if we watch someone travel at a high rate of speed, their clock and their heart will move more slowly than ours. Recalling the story of the Twin Paradox, we found that when Jon traveled at 100,000 miles per second, he aged more slowly than his brother Ron who remained here on Earth. As a matter of fact, we can use the fourth equation of Lorentz to show that the passage of 9.5 minutes for Ron corresponded to the passage of 8 minutes for Jon. First we must rewrite the Lorentz equation to reflect the situation of the twins:

$$\text{Time according to Jon's clock} = \sqrt{1 - \frac{\left(\begin{array}{c}\text{Velocity of Jon}\\ \text{relative to Ron}\end{array}\right)^2}{(\text{Velocity of light})^2}}\left(\begin{array}{c}\text{Time according to Ron's}\\ \text{clock}\end{array}\right)$$

Substituting the proper numbers, we obtain:

$$\text{Time according to Jon's clock} = \sqrt{1 - \frac{(100,000)^2}{(186,000)^2}}\ (9.5)$$

$$\text{Tie according to Jon's clock} = 8 \text{ minutes}$$

If Jon had traveled at 150,000 miles per second, the passage of 9.5 minutes for Ron would have corresponded to the passage of 5.6 minutes for Jon.

$$\text{Time according to Jon's clock} = \sqrt{1 - \frac{\left(\begin{array}{c}\text{Velocity of Jon}\\ \text{relative to Ron}\end{array}\right)^2}{(\text{Velocity of light})^2}}\left(\begin{array}{c}\text{Time according to}\\ \text{Ron's clock}\end{array}\right)$$

$$\text{Time according to Jon's clock} = \sqrt{1 - \frac{(150,000)^2}{(186,000)^2}}\ (9.5)$$

$$\text{Time according to Jon's clock} = 5.6 \text{ minutes}$$

Now if Jon had traveled at the speed of light, 186,000 miles per second, the passage of 9.5 minutes for Ron would have corresponded to the passage of zero minutes for Jon.

$$\text{Time according to Jon's clock} = \sqrt{1 - \frac{\left(\begin{array}{c}\text{Velocity of Jon}\\ \text{relative to Ron}\end{array}\right)^2}{(\text{Velocity of light})^2}}\left(\begin{array}{c}\text{Time according to Ron's}\\ \text{clock}\end{array}\right)$$

$$\text{Time according to Jon's clock} = \sqrt{1 - \frac{(186,000)^2}{(186,000)^2}} \qquad (9.5)$$

$$\text{Time according to Jon's clock} = 0 \text{ minutes}$$

Time would have truly stood still for Jon.

Continuing this process, we find that if Jon had traveled faster than the speed of light, say 200,000 miles per second, the passage of 9.5 minutes for Ron would correspond to the passage of an "imaginary number" of minutes for Jon. An imaginary number is what we get when we try to take the square root of a negative number.

$$\text{Time according to Jon's clock} = \sqrt{1 - \frac{\left(\frac{\text{Velocity of Jon relative to Ron}}{\text{Velocity of light}}\right)^2}{(\text{Velocity of light})^2}} \left(\text{Time according to Ron's clock}\right)$$

$$\text{Time according to Jon's clock} = \sqrt{1 - \frac{(200,000)^2}{(186,000)^2}} \qquad (9.5)$$

$$\text{Time according to Jon's clock} = \sqrt{1 - 1.16} \qquad (9.5)$$

$$\text{Time according to Jon's clock} = \sqrt{-.16} \qquad (9.5)$$

$$\text{Time according to Jon's clock} = \sqrt{.16}\ \sqrt{-1} \qquad (9.5)$$

According to mathematicians,

$$\sqrt{-1} = i$$

which is called the imaginary unit.

Therefore,

$$\text{Time according to Jon's clock} = (.4)\ (i)\ (9.5)$$

$$\text{Time according to Jon's clock} = 3.8\ i \text{ minutes}$$

Since nobody has any idea what $3.8\ i$ minutes really is in reality, it is impossible to reach any conclusion regarding travel at a velocity exceeding the speed of light. However, there is still another good reason which prevents us from traveling faster than the speed of light—a reason which will become apparent in the next chapter.

Chapter 9

# Mass

MASS, LIKE LENGTH AND TIME, UNDERGOES A change for observers who are traveling past each other in a straight line and at a constant velocity. To understand how this change occurs, we must first acquire a somewhat detailed understanding of the concepts of inertia and mass.

A giant meteor flying through space does not encounter air resistance, and will therefore continue traveling in a straight line with uniform velocity unless it encounters some large object or force. A ping pong ball flying through space will do exactly the same (Fig. 9-1). If either the meteor or the ping pong ball encounters an object or force of some significant size, it may be stopped, or its direction of travel may be changed. It seems certain, however, that if the motion of the meteor or ping pong ball is to be affected, the object or force will have to be larger for the meteor than for the ping pong ball. The meteor will offer more resistance than the ping pong ball to any effort made to alter its state of motion. We call this resistance inertia, and we say, in this case, that the meteor has more inertia than the ping pong ball.

Here on Earth, we encounter a similar situation when we consider a locomotive and a ping pong ball which are both standing still. Since it is harder for us to get the locomotive to move than to get the ping pong ball to move, it is obvious that the locomotive offers more resistance than the ping pong ball to any effort made to alter its state of motion.* Once again we call this resistance inertia and say that the locomotive has more inertia than the ping pong ball (Figs. 9-2 and 9-3).

In general, the resistance offered by an object to a force which is trying to start it, stop it, or change its state of motion in any way is called inertia. The concept of inertia, as we will soon see, plays an important role in the definition of mass.

It is possible to think of mass as the 'quantity of matter' in an object—like the number of atoms in an object. The more atoms in an object, the greater its mass. It is important, however, that we do not confuse the concept of mass with the concept of

---

*The phrase 'state of motion' is used even in a situation like this where 'state of stillness' might be more appropriate.

Fig. 9-1. Both a meteor and a ping pong ball will continue traveling through space in a straight line unless deflected or stopped by an outside force.

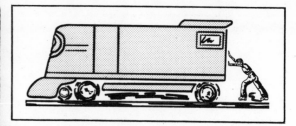

Fig. 9-2. A locomotive offers more resistance than a ping pong ball to any effort made to move it.

weight. The weight of an object is the force with which it is pulled by gravity. The mass of an object (at a given temperature) is the same regardless of where it may be located. In contrast, the weight of an object changes with location. You weigh slightly more at the bottom of a mountain that you do at the top (Fig. 9-4). At the bottom of a mountain, the force of gravity is stronger because you are closer to the center of the Earth. Your mass, however, does not change. The number of atoms in your body is the same whether you are at the bottom of a mountain or the top.

As another example, imagine what would happen if you were to weigh yourself on the Moon. You would find that you weigh roughly one-sixth of what you weigh here on Earth. This is because the gravitational pull of the Moon is roughly one-sixth that of the Earth (Fig. 9-5). Your mass, however, or the number of atoms in your body would still be the same.

To be more specific, we can define mass as a measure of the inertia which an object exhibits in the absence of friction, and in response to any effort made to start it, stop it, or alter its state of motion in any way. Here again, mass does not change with location. Imagine a heavy book sitting on a table top. It takes a certain force, discounting friction, to push the book horizontally across the table. Regardless of where the table is located, at the bottom of a mountain, at the top of a mountain, or on the Moon, it takes the same amount of force to move the book (Fig. 9-6). The difference in gravity at the three locations has no bearing on the amount of force which must be exerted to move the book in a horizontal direction. The book offers the same resistance in all three places, and consequently its inertia and therefore its mass will be the same in all three locations.

Although it appears now that we have two definitions for mass, a little reflection reveals that we merely have two different ways of looking at the same thing. A giant meteor has greater inertia than a

Fig. 9-3. A ping pong ball has less inertia than a locomotive.

Fig. 9-4. You weigh slightly less at the top of a mountain than you do at the bottom.

Fig. 9-5. Your weight on the moon is approximately one-sixth of what it is here on Earth.

Fig. 9-6. Regardless of where you may be located, on the Moon, on the top of a mountain, or at the bottom of a mountain, it takes the same amount of force (discounting friction) to push a book horizontally across a table.

ping pong ball because it offers greater resistance to a force trying to change its state of motion. It also has more atoms, and therefore more mass than a ping pong ball. It follows that the greater the inertia, the greater the 'quantity of matter' in an object, and therefore the greater its mass. In summary, mass is the 'quantity of matter' in an object, or, to be quite specific, it is a measure of the inertia which an object exhibits in the absence of friction, and in response to any effort made to start it, stop it, or alter its state of motion in any way.

Although the mass of an object does not change with location, it does change with velocity. This was first observed in the laboratory, and later explained by Einstein. In 1901, a scientist by the name of Walter Kaufmann found that the mass of a moving electron was greater than the mass of an electron at rest. In 1905, Einstein explained this phenomenon in his paper, "Does the Inertia of a Body Depend Upon Its Energy Content?", the same paper in which he wrote $E = mc^2$. To understand the reason, we must probe still further into the changes which occur in environments that are moving relative to us.

Imagine a rocket which is 10,000 miles long and standing perfectly still. A friend standing at the front of the rocket can use his stopwatch to confirm the fact that a cannonball flying from the back of the rocket to the front at 10,000 miles per hour, will complete the journey in one hour (Fig. 9-7). Now imagine that the rocket is moving across our line of vision in a straight line with uniform velocity. It is

traveling at some high speed close to the speed of light. As far as our friend in the rocket is concerned, everything looks the same as it did before. Since he is unable to detect his motion in space, he still maintains that he is standing still; and the cannonball, speeding along at 10,000 miles per hour, still takes one hour to travel from one end of the rocket to the other. But when we look at the rocket, we notice that it appears to be shorter in the direction of travel, and that our friend's stopwatch moves more slowly than it did when the rocket was standing still. Time passes more slowly in the rocket now that it is moving. The passage of one hour for our friend corresponds to the passage of something greater than one hour for us. Consequently, as far as we are concerned, the cannonball takes longer to travel a distance which is now shorter than 10,000 miles. The cannonball is therefore traveling at some speed which is less than 10,000 miles per hour. In general, we are saying that objects appear to move more slowly in moving environments (Fig. 9-8).

Of course, we reached the same conclusion in our discussion of time. We found that in a moving environment the hands of a clock would move more slowly, hearts would beat more slowly, and people would talk and move around more slowly. We also said that the faster the environment moved, the slower everything would go; and, at the speed of light, time and everything else would stand still. It follows that the cannonball in our rocket will appear to move more and more slowly the faster the rocket

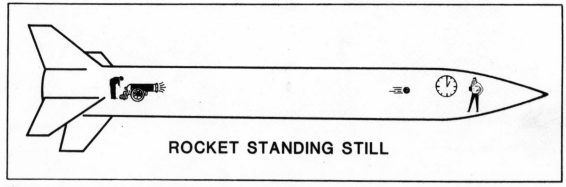

Fig. 9-7. In a rocket which is 10,000 miles long and standing perfectly still, a cannonball flying from the back of the rocket to the front at 10,000 miles per hour will complete the journey in one hour.

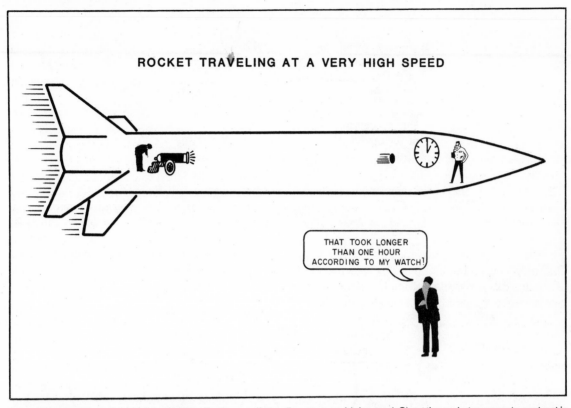

**ROCKET TRAVELING AT A VERY HIGH SPEED**

THAT TOOK LONGER
THAN ONE HOUR
ACCORDING TO MY WATCH!

Fig. 9-8. Imagine a rocket which is 10,000 miles long and traveling at a very high speed. Since the rocket appears to contract in length, we will find that a cannonball flying from the back of the rocket to the front at 10,000 miles per hour takes longer than one hour to travel a distance that appears to be shorter than 10,000 miles. We will therefore conclude that the speed of the cannonball is less than 10,000 miles per hour.

moves, and that it will appear to stop if the rocket reaches the speed of light.

Now consider a situation where the rocket is crossing our line of vision at 15,000 miles per second (Fig. 9-9). Inside the rocket, our friend fires the cannon and thereby imparts a force to the cannonball sufficient to send it flying from one end of the rocket to the other at 25,000 miles per second.

Common sense dictates that we should see the cannonball cross our line of vision at a speed of 40,000 miles per second (15,000 miles per second + 25,000 miles per second). However, as we have already seen, objects in moving environments appear to move more slowly than common sense would dictate. We must use the following formula to calculate the speed of the cannonball:

$$
\text{Our perception of the speed of the cannonball} = \frac{\text{Speed of the rocket} + \text{Our friend's perception of the speed of the cannonball}}{1 + \dfrac{\left(\text{Speed of the rocket}\right)\left(\text{Our friend's perception of the speed of the cannonball}\right)}{(\text{Speed of light})^2}}
$$

Substituting in the proper numbers, we obtain:

$$\text{Our perception of the speed of the cannonball} = \frac{15{,}000 + 25{,}000}{1 + \dfrac{(15{,}000)\,(25{,}000)}{(186{,}000)^2}}$$

$$\text{Our perception of the speed of the cannonball} = 39{,}571 \text{ miles per second}$$

The cannonball, therefore, appears to have increased in speed by 24,571 miles per second (39,571 - 15,000 = 24,571), and not 25,000 miles per second as expected.

Now imagine that the rocket is crossing our line of vision at 150,000 miles per second (Fig. 9-10). Inside the rocket our friend fires the cannon, and once again imparts the same force sufficient to send the cannonball flying from one end of the rocket to the other at 25,000 miles per second. Common sense says that the cannonball will cross our line of vision at 175,000 miles per second (150,000 miles per second + 25,000 miles per second). Again, however, the cannonball will actually move more slowly. Using our formula, we obtain:

$$\text{Our perception of the speed of the cannonball} = \frac{\text{Speed of the rocket} + \text{Our friend's perception of the speed of the cannonball}}{1 + \dfrac{\left(\text{Speed of the rocket}\right)\left(\text{Our friend's perception of the speed of the cannonball}\right)}{(\text{Speed of light})^2}}$$

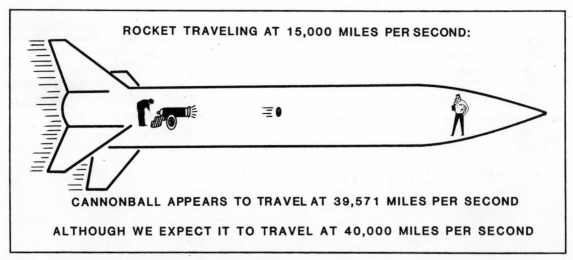

ROCKET TRAVELING AT 15,000 MILES PER SECOND:

CANNONBALL APPEARS TO TRAVEL AT 39,571 MILES PER SECOND

ALTHOUGH WE EXPECT IT TO TRAVEL AT 40,000 MILES PER SECOND

Fig. 9-9. Imagine a rocket crossing our line of vision at 15,000 miles per second. If we fire a cannon and thereby impart a force to a cannonball which moves the ball at 25,000 miles per second, we would expect to see the ball cross our line of vision at 40,000 miles per second. We find, however, that the ball travels at a speed of only 39,571 miles per second.

78

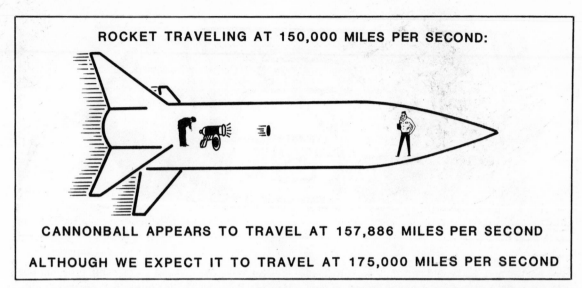

ROCKET TRAVELING AT 150,000 MILES PER SECOND:

CANNONBALL APPEARS TO TRAVEL AT 157,886 MILES PER SECOND

ALTHOUGH WE EXPECT IT TO TRAVEL AT 175,000 MILES PER SECOND

Fig. 9-10. Imagine that the rocket is crossing our line of vision at 150,000 miles per second. If we fire a cannon and thereby impart a force to the cannonball which moves the ball at 25,000 miles per second, we would expect to see the ball cross our line of vision at 175,000 miles per second. We find, however, that the ball travels at a speed of only 157,886 miles per second.

Our perception of the speed of the cannonball $= \dfrac{150{,}000 + 25{,}000}{1 + \dfrac{(150{,}000)\,(25{,}000)}{(186{,}000)^2}}$

Our perception of the speed of the cannonball $= 157{,}886$ miles per second

We see now that the cannonball appears to have increased in speed by only 7,886 miles per second (157,886 − 150,000 = 7,886), and not 25,000 miles per second as expected.

If we were to consider the same set of circumstances without the rocket, we could imagine a ball crossing our line of vision at 15,000 miles per second. If suddenly a force was applied to the ball sufficient to impart an additional speed of 25,000

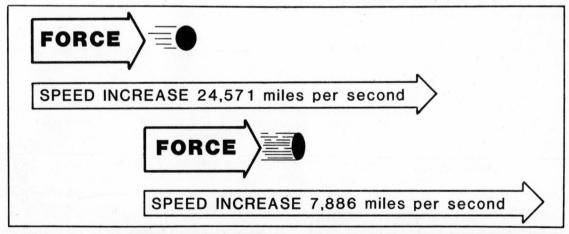

FORCE

SPEED INCREASE 24,571 miles per second

FORCE

SPEED INCREASE 7,886 miles per second

Fig. 9-11. Initially, the ball at the top is traveling at 15,000 miles per second, and the one at the bottom is traveling at 150,000 miles per second. The same force is applied to both balls in an attempt to increase the speed of each by 25,000 miles per second. The illustration highlights the fact that the force is less effective when applied to the faster moving ball.

Fig. 9-12.

miles per second, we would certainly be surprised to find that the ball had increased in speed by only 24,571 miles per second. But, if the same ball was crossing our line of vision at 150,000 miles per second, and the same force was applied to the ball, we would undoubtedly be even more surprised to find that the ball had increased in speed by only 7,886 miles per second. Obviously, the same amount of force is less effective in the second case than in the first, and this of course is due to the initial speed of the ball. The greater the initial speed of the ball, the less effective the force (Fig. 9-11).

If the same force can be less effective the greater the initial speed of the ball, it follows that the faster moving ball is more successful in resisting the force. The faster moving ball has greater inertia because it offers more resistance to a force applied in order to move it. For this very reason, it has greater mass. The increase in mass which we observe in an object which is crossing our line of vision can be computed by a formula which Einstein wrote in his paper in 1905. The formula looks very much like the ones written by Hendrik Lorentz.

$$\text{Mass of object moving across our line of vision} = \frac{\text{Mass of object which is at rest relative to us}}{\sqrt{1 - \dfrac{\left(\dfrac{\text{Velocity of object}}{\text{relative to us}}\right)^2}{(\text{Velocity of light})^2}}}$$

Suppose the mass of a spaceship crossing our line of vision at 15,000 miles per second is two million kilograms* when it is at rest relative to us. Substituting these numbers into our formula, we obtain:

$$\text{Mass of object moving across our line of vision} = \frac{2{,}000{,}000}{\sqrt{1 - \dfrac{(15{,}000)^2}{(186{,}000)^2}}}$$

$$\text{Mass of spaceship moving across our line of vision} = 2{,}006{,}536 \text{ kilograms}$$

If the spaceship is moving across our line of vision at 150,000 miles per second, we can substitute into the formula to obtain:

$$\text{Mass of object moving across our line of vision} = \frac{\text{Mass of object which is at rest relative to us}}{\sqrt{1 - \dfrac{\left(\dfrac{\text{Velocity of object}}{\text{relative to us}}\right)^2}{(\text{Velocity of light})^2}}}$$

*We must use kilograms instead of pounds because kilograms are units of mass whereas pounds are not. In the English system, the unit of mass is called a slug.

$$\text{Mass of object moving across our line of vision} = \frac{2,000,000}{\sqrt{1 - \frac{(150,000)^2}{(186,000)^2}}}$$

$$\text{Mass of object moving across our line of vision} = 3,382,377 \text{ kilograms}$$

Obviously, the faster the spaceship moves, the greater its mass. If it could travel at the speed of light, the formula predicts that the mass of the spaceship would be infinite!

$$\text{Mass of object moving across our line of vision} = \frac{\text{Mass of object which is at rest relative to us}}{\sqrt{1 - \frac{\left(\dfrac{\text{Velocity of object relative to us}}{\text{Velocity of light}}\right)^2}{}}}$$

$$\text{Mass of object moving across our line of vision} = \frac{2,000,000}{\sqrt{1 - \frac{(186,000)^2}{(186,000)^2}}}$$

$$\text{Mass of object moving across our line of vision} = \text{infinity}$$

The fact that the mass of a spaceship would be infinite at the speed of light is another reason why we can never travel at that speed (Fig. 9-12). In order to move a spaceship of infinite mass, we would need an infinite amount of energy; and that is more energy that we have in this entire universe!

# Chapter 10

$$E = mc^2$$

EARLIER IN THIS BOOK, WE INTRODUCED A formula first written by Isaac Newton in the 17th century. The formula is:

Force = Mass × Acceleration

Among other things, the formula tells us that the force you apply when you throw a ball may be determined by multiplying the mass of the ball by its acceleration (Fig. 10-1). As you swing your arm forward and let the ball go, you accelerate the ball from a velocity of 0 miles per hour to perhaps 25 miles per hour at the instant of release. If you exert more effort and apply a greater force, you may succeed in accelerating the ball from a velocity of zero miles per hour to perhaps 35 miles per hour. The more force you apply, the greater the acceleration, or change in velocity of the ball.

Although it was long recognized that a force would accelerate, or change the velocity of a ball, no one ever had any reason to believe that the application of force would change the mass of a ball. After all, why would a moving ball behave as though it had

magically acquired more matter? In 1905, however, Albert Einstein showed that a moving ball would have a greater mass than one that was at rest. This was indeed a remarkable find, because like so many other pronouncements by Einstein, this too flew directly in the face of common sense. To top it off, Einstein used this idea to help him discover one of the most fundamental relationships in nature: the equivalence of mass and energy.

For our purposes, the concept of energy needs no formal definition. The harder a ball is thrown, the more energy it possesses. The greater the force used to throw a ball, the greater its velocity, its mass, and its energy at the instant of release. Intuitively, it is easy to recognize that the faster a ball moves, the more energy it possesses. Although this is indeed true, the additional energy results not only from the increased speed, but from the slight increase in mass as well. Einstein, sensing this relationship between mass and energy, had only to determine how the two were related mathematically. Using the formula he developed which shows the relationship between the mass of an object and

Fig. 10-1. The force used to throw a ball is equal to the mass of the ball times its acceleration.

its velocity, and existing formulas that showed the relationship between the energy and velocity of moving objects, he discovered that mass and energy were related by the speed of light squared (Fig. 10-2). Specifically, the energy possessed by a moving object is equivalent to its mass times the speed of light squared plus the energy of motion. But if the object is not moving, it has no energy of motion, and therefore the relationship is simply $E = mc^2$ (Fig. 10-3). It turns out that mass and energy are really only different forms of the same thing.

We have finally reached the point where we can now answer the question with which we began our story. As the reader will recall, the original question was, "How does starlight manage to travel through the vacuum of empty space?" The answer may be found in the formula $E = mc^2$. Since energy and mass are equivalent, a photon of light, which consists of pure energy, is equivalent to mass. Like mass, it has a reality, or a "substance" all of its own. Consequently, it does not need a medium like the ether in order to travel around the universe. A photon of light behaves like a ping pong ball, and, like a ping pong ball, it does not require a medium in which to travel. What all this amounts to is that if we could weigh a bottle of starlight, it would weigh something (Fig. 10-4). No one doubts that it would be difficult to put starlight into a bottle, and that it

Fig. 10-2.

Fig. 10-3.

would be difficult to find a scale sensitive enough to measure such a trifling amount of weight. But if we could do it, we would not be disappointed.

The formula also tells us much more. For example, if we were to take an atom apart, we would find that it consists of protons, neutrons, and elec-trons. If we were to try to grab one of these parti-cles, such as the electron, we would find ourselves faced with an impossible task. Since it takes a minimum of one photon of light to even see the electron, the photon will move the electron away everytime we attempt to grab it. Consequently, we can never really find the exact location of the elec-tron. The best we can ever say is that there is some probability that the electron may be found here or there, but we can never really pinpoint the exact location (Fig. 10-5). When we attempt to get our hands on hard, substantive, ponderable matter, we find it to be an impossible thing to do. We find that matter, as perceived by our senses, is in reality nothing but a form of concentrated energy just as the formula $E = mc^2$ implies. Just as pure energy, such as starlight, may be perceived as something with "substance", so matter may be perceived as something without "substance". Such are the strange and perplexing results of relativity!

Of course, there is one very dramatic state-ment made by this formula which has come to domi-

Fig. 10-4. If we could weigh a bottle of starlight, we would find that it does indeed weigh something!

Fig. 10-5.

nate modern life. It is the fact that a little bit of mass is equivalent to an enormous amount of energy. $E = mc^2$ explains the release of the tremendous amount of energy liberated in a nuclear fission reaction—a fact which was first dramatically illustrated by a large mushroom shaped cloud that ap-

peared over the sands of the New Mexico desert during the early morning hours of July 16, 1945 (Fig. 10-6).

When Einstein discovered the relationship between mass and energy in 1905, he gave no thought to the possibility that it might one day lead to nuclear weapons. Even as late as 1939, when he signed a letter to President Franklin D. Roosevelt urging him to initiate a study which led to the construction of the atomic bomb, he maintained that a weapon of this sort was unlikely to be built in his lifetime. He never took part in any of the scientific work which led to nuclear fission, and ultimately to an atomic bomb. But he signed the letter to Roosevelt because he shared the concern of his scientific colleagues in America who feared that the Germans might attempt to build such a weapon. Their fears were well founded since it was the Germans, Otto Hahn and Fritz Strassman, who first split the atom in late 1938. Their work was done at the Kaiser Wilhelm Institute in Berlin where Einstein had once been the Director.

On August 6, 1945, the atomic bomb was dropped on Hiroshima. When Einstein was told of the news by his secretary Helen Dukas, he said sadly in German, "Oh weh!" which can only be poorly translated as "Alas!"

Fig. 10-6. The birth of the Atomic Age. For an instant, the light from the explosion was so intense that it could have been seen from another planet (courtesy of Brown Brothers).

# Chapter 11

# What Is Relativity?

UNFORTUNATELY, IT IS VIRTUALLY IMPOSSIBLE to define relativity with a quick, one sentence statement that accurately conveys a meaningful picture. Einstein, himself, formulated the "Principle of Special Relativity" by starting with two facts about the universe, one of which contradicted common sense. By following a very careful line of reasoning, he then proceeded to alter our ideas about length, mass, and time, forming conclusions which also contradicted common sense. One of these conclusions, the variation of mass with velocity, led directly to the discovery of a fundamental relationship in nature: the equivalence of mass and energy. The fact that this formula is accurate is revealed only too dramatically by the existence of nuclear weapons.

Throughout the development of these ideas, Einstein always assumed that observers were moving relative to each other in a straight line and with uniform velocity. In our discussions of length and time, we always assumed that our friends on their Sun-Earth System traveled in a straight line with uniform velocity; and in our discussion of mass, we assumed the same for the travels of the 10,000 mile long rocket. We also found that however fast our friends might be traveling relative to us, they could not detect their motion in space, and so all things appeared to be normal in their own environment. Consequently, any measurements they might make of length, mass, and time would result in the same values we would obtain if we made those same measurements in our own environment. Since length, mass, and time are fundamental units of measure, everything we measure can usually be reduced to these units. In our respective laboratories, therefore, both our friends and we will ultimately discover the same temperature for the freezing point of water, and the same value for the density of mercury or the force of gravity. Whatever we measure, we will both agree on the final value, even though each will claim that the other owns a slow clock and uses a small ruler when measuring things in the direction of travel. Having arrived at the same values for the things we measure, we will ultimately discover that if the same things happen in both environments, they can be explained by the same laws of physics. For example, Newton's laws of motion will apply equally well in both environments, as will the laws which explain the behavior of gases, or the flow of electricity. In general, the "Principle of Special Relativity" holds that:

*If the laws of physics are valid in one environment, they are equally valid in an environment which is moving relative to it in a straight line with constant velocity.*

Everything we have discussed so far falls under the heading of the "Special Theory of Relativity" because in all of our arguments we have assumed that observers are moving relative to each other in a straight line with uniform velocity. Although most people make no distinction between the Special Theory of Relativity and the General Theory of Relativity, it is the results of the Special Theory of Relativity which are so well known. As a matter of fact, it was the Special Theory of Relativity, along with other concepts developed at the turn of the century, which saw the end of the era of classical physics and the beginning of modern physics. It was also the Special Theory of Relativity which taught mankind that never again could he rely solely on his senses to discern the ultimate mysteries of his existence. What we see, hear, taste, touch, and smell is only our individual perception and interpretation of reality. Our environment, it turns out, is not so easily perceived or understood; it is abstract and elusive, perhaps to be explained only by the strange symbols and markings of the mathematician.

For Albert Einstein, the end was not yet in sight. Dissatisfied with the need to assume that observers move relative to each other in a straight line with uniform velocity, he proceeded to "generalize" his theory so that the same conclusions would follow for observers moving relative to each other at any speed and in any direction. This work occupied his time from 1905 until 1916, and resulted not only in a generalized version of the principle of relativity, but in a new law of gravitation as well. Although the results of the General Theory of Relativity are not as well known, their impact upon our conception of the universe is far more dramatic.

**Chapter 12**

# The General Theory of Relativity

IN FORMULATING THE GENERAL THEORY OF RELATIVITY, Einstein began by focusing once again on those two important facts which he had used in building the Special Theory of Relativity.

**Fact One:** It is impossible to detect the motion of the Earth, or any other heavenly body, relative to an ether assumed to be standing perfectly still in the universe. Consequently, it is impossible to know if any heavenly body is truly standing still or moving in the universe.

**Fact Two:** The speed of light is the same regardless of whether the light source is moving or not, and regardless of whether the observer is moving or not.

Einstein reasoned that if he could prove that these two facts were still true in situations where we might be accelerating or changing our direction of travel, then everything that followed from these two facts in the special case would follow in the general case. Before we discuss his reasoning, however, there is one type of motion which warrants special discussion and which will serve to highlight the problem faced by Einstein.

If an object travels in a circle with constant velocity, it is undergoing acceleration. Normally, we think of acceleration as "speeding up" and deceleration as "slowing down." However, physicists think of acceleration as motion accompanied by a force. If we tie a string to a ball and swing it around our head at a constant velocity, we exert a force on the ball through the string (Fig. 12-1). The moment we cut the string, the ball will stop moving in a circle, and the inertia of the ball will cause it to fly off in a straight line moving in the direction it had at the moment we cut the string. As long as we exert a force through the string, the ball will continue to travel in a circle experiencing what physicists call acceleration.

Fig. 12-1. If we tie a string to a ball and swing it around our head, we exert a force on the ball through the string. Even if the ball travels at constant velocity, it experiences what physicists call acceleration.

If we sit in a flying saucer which is accelerating through space, the mere fact that we are pressed against the seat seems to prove that we are subject to a motion which is accompanied by a force (Fig. 12-2). If we can prove that we are in motion, we can successfully contradict Fact One. Therefore, in order to maintain the validity of Fact One, Einstein had to show that it would be impossible for us to detect our motion in space even though we might be accelerating or decelerating, or, to be precise, even though we might be subject to any kind of motion which is accompanied by a force.

Fig. 12-2. If we sit in a flying saucer which is accelerating through space, the fact that we are pressed against the seat seems to prove that we can detect our motion in space.

Fig. 12-3. If we imagine a man in a totally enclosed chest being accelerated upward through space, we will discover that the man finds it impossible to distinguish between gravity and inertia.

Einstein approached this problem by what he called a "thought experiment." He imagined a man located in a totally enclosed chest being accelerated upward through space by some kind of "being." The man is unable to see anything but the inside of the chest. He is being accelerated at a rate which makes him feel as though his feet were being pressed against the floor with a force equivalent to that which he would experience if he was standing on Earth. In other words, the man is having the same sensations as if he was inside a chest which was standing on the ground. In one case, the man is subject to acceleration, and, in the other case, he is subject to the force of gravity. Since he is unable to look outside of the chest in either case, he has no way of knowing if he is moving or standing still.

If we now imagine that the man is holding a ball, and he suddenly opens his hands to let the ball go, we should not be surprised to find that the ball, being subject to inertia, continues moving upward in a straight line at precisely the speed it had the moment the man opened his hand (Fig. 12-3). The chest, however, continues to accelerate, and so the floor of the chest quickly catches up to the ball. The man is now unable to figure out if the floor caught up to the ball, or the ball fell to the floor (Fig. 12-4). In fact, the man is unable to tell if the ball is subject to inertia or gravity. It appears, therefore, as though inertia and gravity are nothing more than two words for the same thing. But if inertia is the same as gravity, Einstein reasoned that it might be possible to write a law of gravity which, unlike Newton's law, did not depend upon a force acting at a distance. In other words, it might be possible to say that we are being held down on this Earth by something other than a mysterious force called gravity. If this law of gravity should prove to be correct, it would prove beyond a shadow of a doubt that inertia and gravity really are only two words for the same thing, and that would in turn prove that it is impossible for us to detect our motion in space even though we may be experiencing motion accompanied by a force.

But what about Fact Two? Is the speed of light the same even though the light source is accelerating or decelerating? Is it the same even though the

Fig. 12-4. If he opens his hand to let a ball go, he will find himself unable to determine if the ball dropped to the floor, or the floor of the chest caught up to the ball.

observer may be accelerating or decelerating? Einstein maintained that the answer was "yes" to both of these questions. If a flying saucer has its landing lights on, and it accelerates or decelerates while crossing our line of vision, both we and the pilot will find that a photon will move away from the ship at 186,000 miles per second (Fig. 12-5). After all, if a photon moves away from the ship at 186,000 miles per second regardless of the ship's cruising speed, there is no reason why the photon should travel at anything other than 186,000 miles per second when the ship is changing speeds.

The same holds true for the case where the light source is stationary and the observer is mov-

ing. If you were a passenger in a flying saucer accelerating or decelerating past a star, you would see a photon move away from the star at 186,000 miles per second (Fig. 12-6). After all, if you see a photon move away from a star at 186,000 miles per second while cruising at any of several different speeds, there is no reason why the photon should travel at anything other than 186,000 miles per second when your ship is changing speeds. Simply stated, the speed of light is constant under any and all circumstances, and therefore Fact Two remains unaltered.

With Fact Two intact, Einstein had only to show that Fact One remained unaltered even in

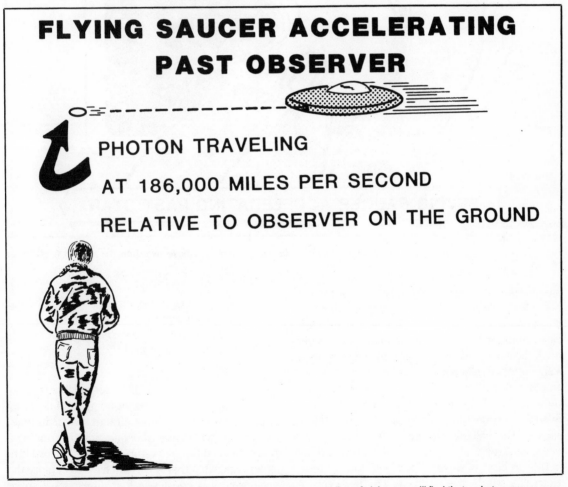

**FLYING SAUCER ACCELERATING PAST OBSERVER**

PHOTON TRAVELING

AT 186,000 MILES PER SECOND

RELATIVE TO OBSERVER ON THE GROUND

Fig. 12-5. If a flying saucer with its landing lights on accelerates past our line of vision, we will find that a photon moves away from the ship at 186,000 miles per second.

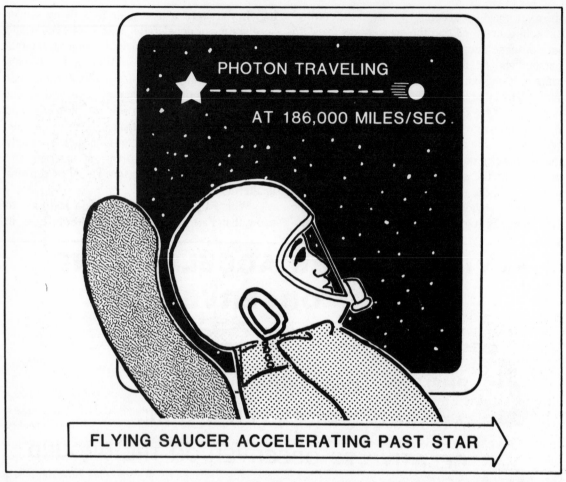

PHOTON TRAVELING

AT 186,000 MILES/SEC.

FLYING SAUCER ACCELERATING PAST STAR

Fig. 12-6. A passenger in a flying saucer accelerating past a star, will see a photon move away from the star at 186,000 miles per second.

situations where motion is subject to a force in order to finally conclude that everything which was true in the special case was also true in the general case. In other words, even though two observers might be accelerating, decelerating, or traveling in a zig-zag or circular path relative to each other, both would agree on the answers they obtained when they measured the same things. In time, both would even develop the same laws of physics, even though each would maintain that the other owns a slow clock and uses a short ruler when measuring things in the direction of travel. The "Principle of Special Relativity" could then be changed to the "Principle of General Relativity":

*If the laws of physics are valid in one environment, they are equally valid in an environment which is moving relative to it.*

It was quite obvious that a formidable task lay ahead. After all, if we are not being held down by gravity, why don't we float off into space? In time, Einstein would answer this question by creating a concept called the "curvature of space." In order to understand what this is, we will have to explore the mysteries of geodesics in four-dimensional space time.

Chapter 13

# The Shape of Things

WE LIVE IN A WORLD OF FOUR DIMENSIONS; three dimensions of space (length, width, and height) and one dimension of time. When we stand still, we stand still in space, at least relative to the Earth, but move forward in time. If we look at the stars in the sky, they appear to be "fixed" in space even though we know that they appear in space, we see them all at different points in time because it takes years for the light from the stars to reach us here on Earth. We see some stars as they appeared 50 years ago, others as they appeared 100 years ago, and still others as they appeared 2000 years ago. Their age depends upon how far away they are from us. Four dimensional space-time is really nothing more than a composite of three dimensions of space and one dimension of time. Although we live in four dimensions of space and time, it is important to realize that we can only see in the the three dimensions of space.

A geodesic is the path followed by a heavenly body. A heavenly body, like everything else in the universe, exists in both space and time, and so does

the path traced by a heavenly body. A geodesic in space is the shortest distance between two points in that space. A geodesic on this page is a straight line because the shortest distance between two points on a flat, two-dimensional space is a straight line (Fig. 13-1). A geodesic in your living room is a straight line because the shortest distance between two points in a three-dimensional space is a straight line (Fig. 13-2). However, a geodesic on the surface of this Earth is the arc of a circle because the shortest distance between two points on the surface of a sphere is an arc of a circle. If you travel from New York to London, you will traverse the arc of a circle if you hope to cover the shortest possible distance (Fig. 13-3). Again, a geodesic in space is the shortest possible distance between two points in that space.

A geodesic in time is the path corresponding to the passage of the greatest amount of time. Whenever we think of space travel, we typically think of high speeds. Heavenly bodies, however, travel along a geodesic in time by moving at the

# GEODESIC

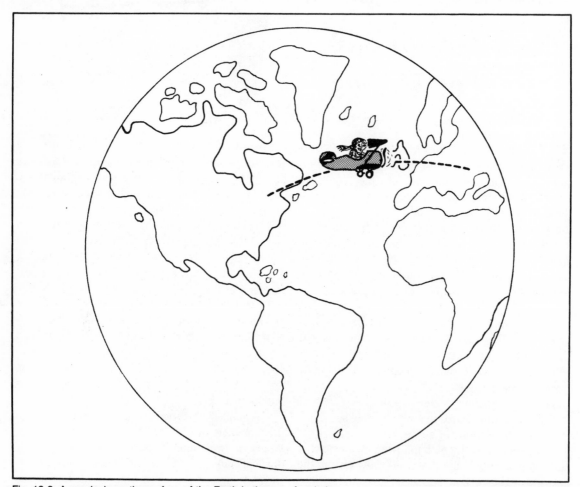

Fig. 13-1. A geodesic in flat, two-dimensional space is a straight line.

slowest possible speed. A comet falling through the heavens, for example, follows a geodesic in time by traveling at the slowest possible speed. Since the comet is moving as slowly as possible, the hands of a clock located on the comet will turn as quickly as possible, thereby recording the passage of the greatest amount of time (Fig. 13-4).

Fig. 13-2. A geodesic in your living room is a straight line.

Fig. 13-3. A geodesic on the surface of the Earth is the arc of a circle.

Fig. 13-4. The hands of a clock located on a comet move as quickly as possible because the comet moves through space as slowly as possible. (Commenting on this phenomenon, the British mathematician and philosopher, Bertrand Russell, once said that a law of "cosmic laziness" prevails throughout the universe.)

Fig. 13-5. If you stand across the street from a man dropping a ball from the roof of a tall building, you will notice that the ball traces a straight line as it falls to the ground.

What is so important about a geodesic that requires all this discussion? Imagine a man dropping a ball from the roof of a tall building (Fig. 13-5). If you face the building as he drops the ball, you will notice that the ball traces a straight line as it falls to the ground. Now imagine that you are standing on the Sun. Since the Earth is now crossing your line of vision, the ball traces the arc of a circle as it falls to the ground (Fig. 13-6). If the Earth stood still relative to the Sun, the ball would once again trace a

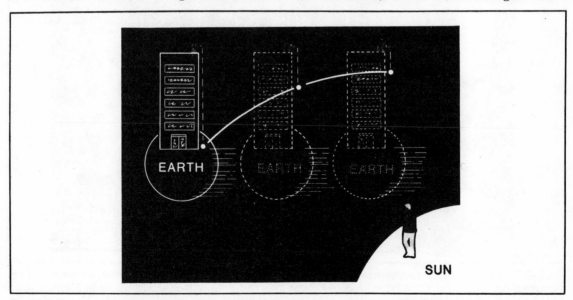

Fig. 13-6. However, if you stand on the Sun, the ball traces the arc of a circle as it falls to the ground.

straight line as it fell to the ground. But now imagine that you are standing on some distant star. Both the Sun and the Earth are now crossing your line of vision relative to the star. Once again, the ball traces the arc of a circle as it falls to the ground.

Since there is not a corner or crevice in this universe where you can possibly stand with the assurance that you are standing still relative to all creation, you can never be certain of holding the Earth perfectly still (Fig. 13-7). Consequently, the ball

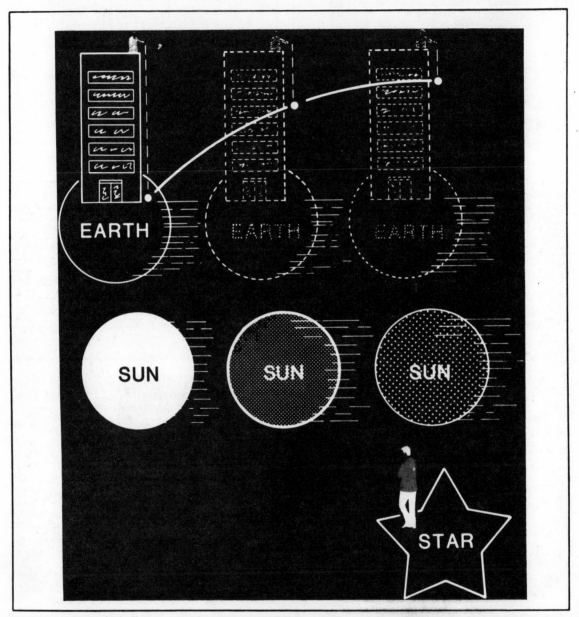

Fig. 13-7. If you stand on a star, or any other place in the universe, the ball traces the arc of a circle as it falls to the ground. Since you cannot stand still relative to all creation, the ball will always trace the arc of a circle. In an absolute sense, it is impossible to trace a straight line in this universe.

Fig. 13-8. When we look at our shadow on the ground, we see a two-dimensional reflection of a three-dimensional reality.

will always trace the arc of a circle. Relative to all creation, it is simply impossible to trace a straight line in this universe—drunk or sober!

What then are we looking at when we notice a comet falling through the heavens? It turns out that the comet traces a path which is really a geodesic in four dimensional space-time. But, remember, we can only see in three dimensions of space. When we look at our shadow on the ground, we see a two-dimensional reflection of a three dimensional reality (Fig. 13-8). When we look at the path of a comet, we see a three dimensional reflection of a four dimensional reality—a geodesic in four dimensional space-time.

We now have all of the essentials necessary to

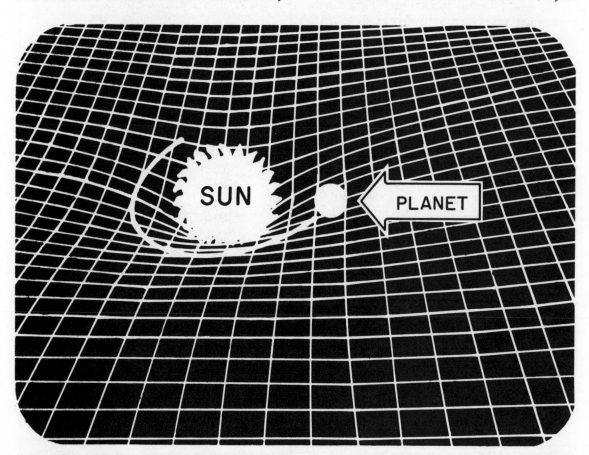

Fig. 13-9. The shape of the universe is analogous to a large rubber sheet punctuated by the presence of large balls which form depressions in the sheet. If we imagine that one of these large balls is the Sun, and a small marble rolling along the rubber sheet is a planet, then we will find that the "planet" circles the "Sun" because it gets trapped in the depression caused by the "Sun". The "Sun" does not exert some mysterious force at a distance called gravity.

form a "picture" of space as Einstein saw it. Just as we may think of a painting as a composite of individual brush strokes, so may we think of space as a composite of individual geodesics. We have only to go outside and record the geodesics of comets, planets, meteors, stars, and other heavenly bodies, and then put them together to obtain a "picture" of space. This composite of geodesics is perhaps best visualized as a large rubber sheet. To get a complete "picture" of the universe, we have only to place a ball on the rubber sheet which would represent the Sun (Fig. 13-9). The ball, of course, creates a depression in the rubber sheet. If we now think of a marble, which might represent a planet, rolling along the rubber sheet, we will find that the marble circles around the ball only if it gets trapped in the depression caused by the presence of the ball. The ball does not influence the marble by exerting some mysterious force at a distance; instead, the "curvature of space" in the vicinity of the ball causes the "planet" to circle the "Sun."

Of course, our "picture" of the universe is far from perfect. Suppose we placed several balls on the rubber sheet, and each of the balls represented a star. Furthermore, suppose our marble represented a comet. If we rolled the marble along the rubber sheet, it would trace a straight line as long as it failed to encounter a depression caused by one of the balls. But, as we said earlier, it is impossible to trace a straight line in this universe. Only by standing still relative to all creation can we accurately depict the "comet" tracing a straight line. Our "picture" also fails to depict the fourth dimension of time. Despite these and other imperfections, our "picture" conveys the main point: heavenly bodies merely follow the curvature of space as they wander about the universe. They are not subject to a force of gravity, because in Einstein's theory there is no such thing as a force of gravity.*

We can now answer the question posed in the last chapter, "What is holding us down on this

---

*Although Einstein replaced the force of gravity with the curvature of space, we still continue to use the concept of gravity. Gravity is a concept which is easy to comprehend, and scientists have always known that calculations based on this concept are sufficiently accurate for most practical purposes.

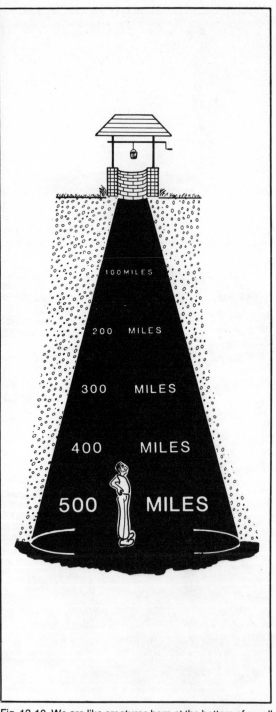

Fig. 13-10. We are like creatures born at the bottom of a well which is hundreds of miles deep. However, the walls of our well are not made of bricks and mortar but of space.

Fig. 13-11. Imagine yourself alone in empty space—just you and black eternity. Are you in the middle of empty space, or are you in the middle of nothing?

Earth?" The answer is that there is nothing pulling us down and nothing pushing us down. We are like creatures born at the bottom of a well which is hundreds of miles deep (Fig. 13-10). The walls of our well are not made of bricks and mortar but of space. This space, in the neighborhood of our Earth, is sufficiently curved so that we "sense" a space which is vertical, like the walls of a well. And just the way it would take an enormous amount of energy to get out of a real well which is hundreds of miles deep, so it takes an enormous amount of energy to get out of the well in space in which we find ourselves. It is no wonder that we require powerful rockets capable of accelerating us to a velocity of 25,000 miles per hour in order to leave our planet Earth.

How does all this fit into common sense? After all, when we look at the heavens, most of what we see is really nothing more than empty space. Imagine yourself alone in that empty space—no stars, planets, comets, or meteors—just you and black eternity (Fig. 13-11). Are you in the middle of empty space or are you in the middle of nothing? According to the General Theory of Relativity, you are in the middle of empty space because it is impossible for you or anything else to exist in the middle of nothing. The implication is that empty space is more than nothing. In a sense, empty space has "substance"; it can be curved, it can be warped, it can and does have "shape". It can even be described by equations written by Albert Einstein.

**Chapter 14**

# The Proof
# of The Pudding

EINSTEIN'S PAPER ON THE GENERAL THEORY OF Relativity was published in 1916, and is considered to be a towering monument of scientific thought. Most scientists agree that had Einstein not published the Special Theory of Relativity when he did, someone else would have published it not long afterwards. However, they are also quick to point out that had Einstein not published the General Theory of Relativity, it might still have remained unknown to this day. He was truly alone when he developed this theory.

In attempting to prove his General Theory, Einstein addressed a question which had been troubling astronomers for years. Recall that Johannes Kepler said that the planets travel around the Sun in elliptical orbits. Years later Isaac Newton introduced his Law of Universal Gravitation and showed that each planet actually shifted into another elliptical orbit every time it circled the Sun. The set of ellipses thus formed create a rosetta pattern. Newton explained that this was the result of the influence of each planet upon the other. His equations even predict the amount of shift experi-

enced by each planet with a fair degree of accuracy. However, such was not the case for the planet Mercury. Since the middle of the nineteenth century, astronomers had realized that the shift experienced by the planet Mercury was considerably greater than that predicted by Newton (Fig. 14-1). To account for this fact, an astronomer by the name of Jean Joseph Leverrier suggested that there might possibly be another, as yet undiscovered planet, exerting a pull on Mercury. Leverrier named the planet Vulcan, and suggested that it might be located on the opposite side of the Sun, moving at a speed comparable to the Earth, so that at best it would be difficult to find. For the next several decades, astronomers tried to find Vulcan by searching for it in the vicinity of the Sun during solar eclipses. Despite repeated attempts, their efforts failed (Fig. 14-2).

In his 1916 paper, Einstein showed that the path of Mercury could be explained by the curvature of space in the vicinity of the Sun. Since Mercury was closest to the Sun, it encountered more curvature than any of the other planets, and, not surpris-

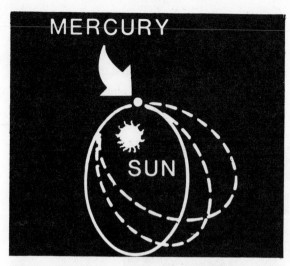

Fig. 14-1. Since Mercury is the closest planet to the Sun, it is most strongly influenced by the curvature of space in the vicinity of the Sun. Each orbit of Mercury is offset from its preceding one by an amount greater than that found for any other planet. The General Theory of Relativity was the first to correctly explain this phenomenon.

ingly, its orbit was most influenced by that curvature. Using his newly developed equations, Einstein was able to show that each successive orbit of Mercury was offset from the preceding one by an amount completely consistent with astronomical observations. To Einstein, this calculation was the first proof of the General Theory of Relativity, and he later told a friend " . . . I was speechless for several days with excitement."

The scientific world accepted Einstein's account of the behavior of Mercury's orbit as only one of several possible explanations of the phenomenon. At the time, it was not generally accepted as definitive proof of his General Theory. The first experimental evidence which was in fact accepted as proof appeared in the year 1919. It was a proof which Einstein himself suggested, and which involved a phenomenon that no one had ever even thought possible—the bending of light. To understand the experiment, let us return to the man in the chest being accelerated through space. Remember that everything that happens in this chest also happens in a chest which is standing on the ground. For example, as the chest accelerates and attains greater and greater speed, a clock located inside the chest will move more and more slowly, and the man inside the chest will age more and more slowly. But the man will be unable to determine if he is being subject to greater acceleration or to a larger force of gravity. It follows that if the man chooses to live on a planet which has an extremely large gravitational pull, he will age more slowly. In fact, anyone who wishes to live a very long life has only to find a planet with a suitable atmosphere, ample vegetation, friendly neighbors, and a very large gravitational pull! As another example, imagine what the man inside the moving chest would see if a photon happened to enter from one side (Fig. 14-3). As the photon crosses to the other side of the chest, the man would watch it fall to the floor (Fig. 14-4). If the chest happened to be located on a planet with a large gravitational pull, the man would conclude that the

Fig. 14-2. Although astronomers never found the planet Vulcan in their telescopes, it did eventually appear on our television and movie screens. Mr. Spock of "Star Trek" comes from Vulcan (courtesy of Wide World Photos).

photon fell to the floor because of the presence of gravity.

Einstein reasoned that if photons could be influenced by the force of gravity, then photons traveling close to the Sun would be attracted by the Sun. Having already replaced the concept of gravity with the curvature of space, Einstein went on to suggest that photons would really follow the curvature of space in their travels past the Sun. To prove that this was true, he suggested that stars near the perimeter of the Sun be photographed during a solar eclipse (Fig. 14-5). He predicted that scientists

Fig. 14-3. To a man located in a totally enclosed chest which is being accelerated upward through space, a photon entering one side of the chest will appear to fall to the floor.

Fig. 14-4. Since the man is experiencing the same sensations as he would if the chest were located on a planet with a large gravitational force, he may very well conclude that the photon fell to the floor because of the pull exerted by gravity.

would find that these stars were further away from the Sun than they had anticipated, because they would be looking up a shaft of light, which, in following the curvature of space, approached the Earth at a slightly steeper than normal angle. He calculated the amount by which the angle would increase so that the observed results could be compared to his predicted results. In the course of these calculations, he also determined that his predicted angle was twice as large as the angle obtained assuming photons were attracted only by the gravity of the Sun and there was no such thing as the curvature of space. If his theory was correct, the angle measured by scientists would equal his predicted angle, and there would be little doubt that Newton's idea of gravity was only an approximation of Einstein's more rigorous concept of the curvature of space.

On May 29, 1919, an eclipse of the Sun took place which could best be observed from South America and Africa. Despite the fact that Britain was in the middle of a war with Germany, the necessary funds were provided to the Royal Society and Royal Astronomical Society establishing a Joint Permanent Eclipse Committee charged with making the necessary measurements. The Committee sent out two expeditions: one to Sobral in northern Brazil, and another to Principe Island in the Gulf of Guinea. Once the photographs were taken, it took the entire summer to develop the plates, perform the measurements, and complete the calculations. The results began to filter out slowly; first only to those connected with the expeditions, then to a handful of scientists, and finally to the world.

When Einstein first heard the news, it came in the form of a telegram from H.A. Lorentz. That same day, September 27, 1919, he wrote to his mother, "Good news today, H.A. Lorentz has wired me that the British expeditions have actually proved the light shift near the sun." A little more than a month later, on Thursday, November 6, 1919, the Fellows of the Royal and Royal Astronomical Societies met in London to hear the official announcement of the results of the two expeditions. The tone of the meeting was perhaps best described by the mathematician and philosopher, Alfred North Whitehead:

". . . The whole atmosphere of tense interest was exactly that of the Greek drama: we were the chorus commenting on the decree of destiny as disclosed in the development of a supreme incident. There was dramatic quality in the very staging—the traditional ceremonial, and in the background the picture of Newton to remind us that the greatest of sci-

Fig. 14-5. Einstein predicted that during a solar eclipse stars near the perimeter of the Sun would appear to be farther away from the perimeter than we might normally expect. In this illustration, we see light from such a star being bent towards the Sun, and we find ourselves looking up a shaft of light which makes the star appear to be "higher" up in the heavens, and, consequently farther away from the perimeter than it really is.

Fig. 14-6.

entific generalizations was now, after more than two centuries, to receive its first modification. Nor was the personal interest wanting; a great adventure in thought had at length come safe to shore."

When Sir Joseph Thomson, president of the Royal Society spoke, he called the results "one of the greatest—perhaps the greatest—of achievements in the history of human thought . . . It is not the discovery of an outlying island but of a whole continent of new scientific ideas. It is the greatest discovery in connection with gravitation since Newton enunciated his principles."

Years earlier, Max Planck, a great physicist in his own right had said, "If Einstein's theory should prove correct, as I suspect it will, he will be considered the Copernicus of the twentieth century." The next morning, November 7, 1919, Albert Einstein awoke in Berlin to face a new life—the one foretold by Max Planck.

**Chapter 15**

# The
# Outer Reaches

Einstein's work on general relativity led to speculations about the size and shape of the universe. As a matter of fact, it was Einstein himself who started all the speculations with a paper published in 1917 titled, "Cosmological Considerations on the General Theory of Relativity". In many respects, scientists today consider this paper to be both outdated and incorrect. But whatever its shortcomings, it remains an historical milestone because most scientists agree that its appearance marked the beginning of modern cosmology. Cosmology is the branch of science concerned with the study of the size, shape, and age of the universe, and its changing features over time. It is a curious blend of theory, speculation, and astronomical observations, as fascinating as it is frustrating.

Typically, when scientists discuss the size and shape of the universe, interest centers on whether the universe is bounded or unbounded, and finite or infinite. The Earth is both unbounded and finite. It is unbounded because you cannot fall off any end, and it is finite because there is only a limited amount of land and water. Isaac Newton thought that the uni-

verse was bounded and finite. But he imagined that all of the heavenly bodies in the universe were clustered in the center of a vast expanse of emptiness (Fig. 15-1). For Newton, the ends, or boundaries, of the universe occurred where the heavenly bodies stopped and the vast expanse of emptiness began. Furthermore, there were just so many heavenly bodies—a finite number. For Gottfried Wilhelm Leibniz, a philosopher and mathematician as well as a contemporary of Isaac Newton, the universe was both unbounded and infinite. Leibniz believed that an infinite number of heavenly bodies were uniformly distributed throughout an infinitely large space (Fig. 15-2). In 1917, Albert Einstein described a universe which was unbounded and finite, just like the Earth. But before we can understand why Einstein's universe was unbounded and finite, we must first try to picture it in our minds. Once again, we must resort to an analogy.

If we were to see a film of a person sitting at a table and peeling an orange, we might witness a scene where the person pulls one piece of skin after another off the orange, and puts each piece down on

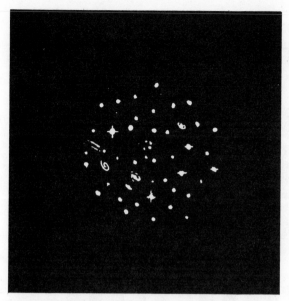

Fig. 15-1. Isaac Newton thought that all heavenly bodies were clustered in the center of a vast expanse of emptiness.

the table in front of him. Now if we were to run the film backwards, we would watch the person take the last piece of skin and put it back on the orange. We would then watch as another piece of skin was picked up and put back on the orange. As this piece of skin was put next to the one already on the orange, the edge between the two pieces would disappear. Next, a third piece of skin would be put back on the orange, and again the edge between this piece and the attached skin would disappear. As we reach the beginning of the film, all of the skin would be put back on the orange, and all of the edges will have disappeared (Fig. 15-3).

Instead of pieces of skin, now imagine that you have a collection of "volumes" such as cubes. Pick up two cubes, put them together, and the surface between them disappears. Pick up a third cube and put it next to the first two, and again a surface disappears. Continue this process of putting cube next to cube, and each time a surface disappears (Fig. 15-4). Finally, try to imagine piling cube on top of cube until even the outside surface disappears! Obviously, this is asking too much. It is impossible to imagine a cube without an outside surface. But we can easily imagine what we would see if we were

inside one of the cubes which made up this enormous pile. We would simply find ourselves in a three dimensional environment; the very same environment we live in everyday.

Einstein did not think that the universe was shaped like a cube without an outside surface, but he did think that it was shaped like a sphere without an outside surface. If we had a collection of hemispheres which we could put together to form spheres, and if each sphere formed in this way was slightly larger than the previous one, we could form a collection of concentric spheres by enclosing each sphere in a larger one (Fig. 15-5). Each time we enclosed a sphere with a larger one, a surface would disappear. If we could continue the process until even the outside surface of the largest sphere disappeared, we would have a pretty fair representation of the universe as Einstein saw it in 1917.

Needless to say, Einstein did not forget to include time in his description of the universe. His equations revealed that the universe was ageless; it never had a beginning, and it would never have an end—it simply had existed forever, and it would continue to exist forever.

Fig 15-2. Gottfried Wilhelm Leibniz thought that an infinite number of heavenly bodies were distributed throughout an infinitely large space.

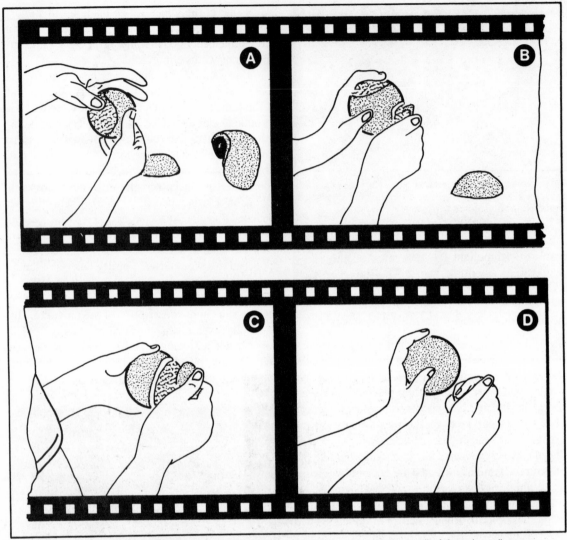

Fig. 15-3. Rolling the film backwards, we find that as the skin is put back on the orange all of the edges disappear.

Having described Einstein's universe, let us see what happens when we explore it. If we were to leave the Earth and travel straight out into space in a rocket moving at some speed very close to the speed of light, we would inevitably travel in a huge circle and ultimately return to the Earth (Figs. 15-6 and 15-7). Throughout the entire journey, we would have no reason to believe that we were traveling in any other way than in a straight line. Our trip would be very similar to a journey around the world in which we appear to travel in a straight line, but

always return to our starting point. Since our space trip unavoidably leads back to our starting point, Einstein's universe is said to be boundless. Despite every effort to fly straight ahead, we never seem to find an end to this universe. Furthermore, Einstein's universe is said to be finite because regardless of how long we spend flying around this universe, we can only pass a certain number of heavenly bodies contained in a very large, but finite volume of space.

Einstein's model of the universe did not re-

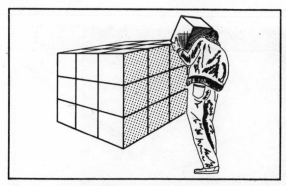

Fig. 15-4. Put cube next to cube and each time a surface disappears. Now imagine that we continue this process until even the outside surface disappears!

Fig. 15-5. As we enclose one sphere inside of another a surface disappears. If we could continue this process until even the outside surface of the largest sphere disappears, we would be left with a fairly accurate idea of Einstein's conception of the universe as he saw it in 1917.

main in the limelight for very long. Within a few years, other scientists had delved into the equations of General Relativity and came up with entirely different conceptions of the universe. In 1929, the astronomer Edwin Powell Hubble made a discovery which proved conclusively that Einstein's

model of the universe was incorrect in several respects. Early in the twentieth century, astronomers had come to realize that the universe consisted of galaxies, (groups of millions of stars) moving through vast reaches of empty space. Hubble showed that all of these galaxies (with the excep-

Fig. 15-6. A Journey Through Einstein's Universe.

Fig. 15-7. If we were to leave the Earth and travel straight out into space, we would ultimately return to our starting point even though we had traveled in a straight line during the entire journey.

tion of those closest to us) were moving away from us and from each other at enormous speeds (Fig. 15-8). In fact, the further away the galaxy, the faster it receded into the distance. With the space between galaxies increasing, it was suggested that perhaps the universe was expanding. If it was expanding, then the beginning of the expansion, or the "big bang" was an event that must have occurred at some particular point in time. Compared to Einstein's universe which was static, this expanding universe was dynamic; and time was not infinite, but began with the "big band". Faced with these ideas, some scientists have even suggested, and in fact many believe, that perhaps the universe pulsates, undergoing expansion and contraction in an endless repetitive cycle.

To this day, theories about the size, shape, and age of the universe abound, but the ultimate answers remain as elusive as ever. To add to the frustration, we now know that there are certain "ultimate" answers which may forever remain beyond our reach. At the present time, for example, it seems as though we will never know how fast the Earth is moving relative to all creation, nor will we ever be able to pinpoint the exact location of an electron at a given moment in time. Of course, there are numerous other examples of this sort, and they all highlight the fact that scientists have raised questions which may never be answered. Hopefully, cosmologists will one day discover the size, shape, and age of the universe; but not withstanding their efforts, there is still the possibility that we may one day have to accept the fact that there are no answers, but only questions!

Fig. 15-8. Hubble proved that all galaxies, with the exception of those closest to us, are moving away from us and from each other. It appears, therefore, that the universe is expanding.

# The Ultimate Mysteries

**Chapter 16**

DESPITE THE INEVITABLE INTERRUPTIONS AND distractions which became a part of his life as a world famous scientist, Einstein continued to pursue his scientific work with a passion and devotion that proved to be more and more remarkable as the years passed.

After achieving international recognition in 1919, he was drawn into the Zionist Movement, and later persecuted by the Nazis and forced to leave Germany. In the autumn of 1933, he emigrated to the United States where he remained for the rest of his life. He lived in Princeton, New Jersey working at the Institute for Advanced Study which is part of Princeton University. Eventually, Einstein became an American citizen, and played a small but significant role in the events leading to the construction of the atomic bomb. Toward the end of his life, he was even asked to be President of the state of Israel. But whatever the demands or pressures, he always managed to return to his first love—theoretical physics (Fig. 16-1).

During these later years, he continued work-

ing on the General Theory of Relativity and made significant improvements in at least two areas. As we now know, when Einstein first published his Theory of General Relativity in 1916, he successfully explained the motion of the planet Mercury. But in doing so, he had to assume that Mercury was a point in space while the Sun was a huge massive body. His equations were not yet sufficiently refined to allow him to predict the motion of two massive bodies revolving around each other—like a binary star system. By 1938 he did manage to solve this problem.

He worked on another improvement in the General Theory during and immediately following the Second World War. In 1916, the equations for the curvature of space were distinctly different from those for the geodesic (path followed by a heavenly body). Working in collaboration with the physicist Leopold Infeld, Einstein was able to refine the mathematics so that the equations for the geodesic could be derived from the equations for the curvature of space. In this sense, the General

Fig. 16-1. Albert.Einstein (circa 1953). Einstein was photographed so much that when a stranger once asked his calling, he replied, "I am a model " (courtesy of Brown Brothers).

Theory of Relativity became strictly a field theory; a theory based only on the shape or geometry of space.

For the last 40 years of his life, Einstein's overriding passion was his pursuit of a Unified Field Theory—a theory which he hoped would unify all of the laws of nature. In the nineteenth century, James Clerk Maxwell wrote a set of equations which brought electricity and magnetism together so that today scientists can deal comfortably with the concept of an electromagnetic field—a combination of an electric field and a magnetic field. Einstein hoped to develop a theory which would combine an electric field, a magnetic field, and a gravitational field thereby explaining phenomena at both a micro and macro level. The equations that would describe the motion of an electron traveling around the nucleus of an atom, would also describe the motion of the Earth traveling around the Sun.

The implications of developing such a theory are mind boggling. A thorough understanding of fields, for example, might eventually lead to anti-gravity devices. The day may come when we will travel around in a car which hovers over the ground by creating an anti-gravity field; or perhaps we may travel to other planets in a rocket which repels (rather than propels) itself away from the Earth (Fig. 16-2). Or, we might even learn how to control the motion of heavenly bodies, and move planets and stars around the universe at will. The machines emerging from a Unified Field Theory would undoubtedly be every bit as miraculous, and hopefully not as sinister, as those fathered by the Theory of Relativity.

Albert Einstein died on April 18, 1955 at the age of 76. Death came in the early morning hours in Princeton Hospital where he had been a patient for several days. A rupture of the aorta, the main artery of the body, was the cause of death. Lying on his hospital night table were papers filled with his latest Field Theory calculations. As he had promised, he worked to the very end. To this day scientists are still trying to complete this difficult work (Fig. 16-3).

Einstein's ideas have become more and more relevant in the years since his death. The space age has opened new vistas and new frontiers. All sorts of strange phenomena and objects have been discovered by astrophysicists, some of which could even be predicted by the General Theory of Relativity.

One of the most interesting of these discoveries is the black hole: a star in which matter has collapsed making it an extremely dense object, and at the same time making it an object with such a strong gravitational field that not even light can

Fig. 16-2. One day we may travel in a car which hovers over the ground by creating an anti-gravity field.

Fig. 16-3. This cartoon appeared shortly after Albert Einstein's death (courtesy of Herblock Cartoons).

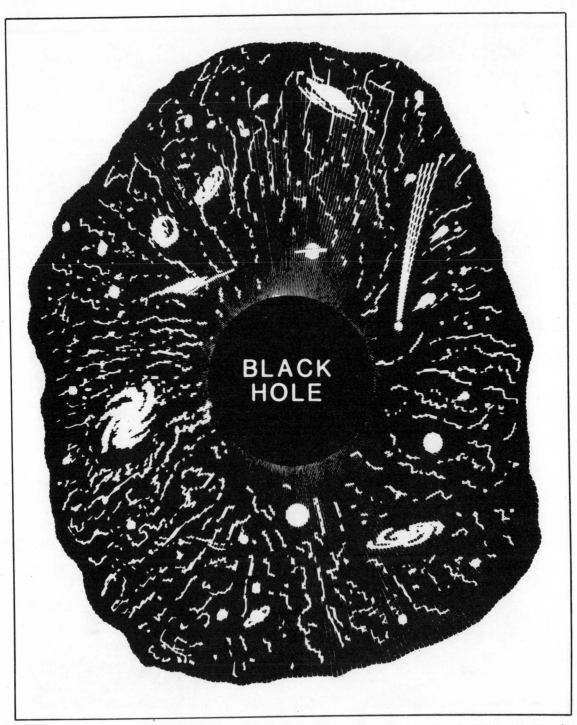

Fig. 16-4. A black hole. An extremely dense object with such a strong gravitational field that not even light can escape its surface.

escape its surface (Fig. 16-4). We have here the ultimate light bending experiment; an object which will not even let light skim across its surface without bending it to the point where it captures it completely. We also have here an object with such a strong gravitational field that only the General Theory of Relativity can describe it; Newton's law of gravity will simply not suffice. Along these same lines, astrophysicists have also used some of their new discoveries to test and retest the validity of the Theory of Relativity. So far it appears to be every bit as valid as Albert Einstein thought it was.

Throughout the world the search continues; scientists everywhere are still trying to unravel the ultimate secrets of nature. Experiments are performed, new theories are propounded, and revolutions in thought take place. The process is slow and tedious, but in the end each vision is replaced by a broader vision. Isaac Newton once wrote, "If I have seen further, it is by standing upon the shoulders of Giants". Today we stand upon the shoulders of Albert Einstein. Whether or not we will ever answer the ultimate questions remains to be seen, but so far we have at least learned something about ourselves. To appreciate that lesson, consider the known size of the universe. If we take the distance from our Earth to the most distant object we can find in the heavens, we can imagine that distance to be the radius of a large sphere. If we now shrink that giant sphere down to the size of our planet Earth, the Earth will in turn shrink to a size of less than one atom; that is how infinitesimally small we humans are in the entire scheme of things. But despite our small size, the human mind, as exemplified by Albert Einstein, is capable of assimilating and comprehending all of the vast surrounding beauty and harmony; and discerning the riddles of the universe, as Albert Einstein liked to call them; or, to a paraphrase Sir Winston Churchill, the secrets of the universe, which are wrapped in mysteries, which are found inside of enigmas!

# Bibliography

## THEORY OF RELATIVITY

Barnett, Lincoln 1948. *The universe and Dr. Einstein*. New York: William Sloan Associates.

Born, Max 1962. *Einstein's theory of relativity*. New York: Dover Publications, Inc.

Chester, Michael 1967. *Relativity: an introduction for young readers*. New York: W.W. Norton & Co., Inc.

Coleman, James A. 1954. *Relativity for the layman*. New York: The MacMillan Company.

Durell, Clement V. 1924. *Readable relativity*. New York: Harper & Row, Publishers.

Einstein, Albert 1916. *Relativity*. New York: Crown Publishers, Inc.

Einstein, Albert 1916. *Relativity: the special and the general theory*. New York: Crown Publishers, Inc.

Einstein, A., Lorentz, H.A., Minkowski H., and Weyl, H. 1923. *The principle of relativity: a collection of original memoirs on the special and general theory of relativity*. New York: Dover Publications, Inc.

Gardner, Martin 1962. *Relativity for the million*. New York: The MacMillan Company.

Gibilisco, Stan 1983. *Understanding Einstein's theories of relativity—man's new perspective on the cosmos*. Blue Ridge Summit, PA: TAB BOOKS Inc.

Hawking S.W. and Israel, W. 1979. *General relativity: An Einstein centenary survey*. London: Cambridge University Press.

Kondo, Herbert 1966. *Adventures in space and time*. New York: Holiday House.

Lanczos, Cornelius 1965. *Albert Einstein and the cosmic world order*. New York: John Wiley & Sons, Inc.

Lieber, Lillian R. 1945. *The Einstein theory of relativity*. New York: Rinehart and Co., Inc.

Mermin, N. David 1968. *Space and time in special relativity*. New York: McGraw-Hill Book Company.

Russell, Bertrand 1925. *ABC of relativity*. London: George Allen & Unwin Ltd.

Struble, Mitchell 1973. *Web of space-time*. Philadelphia: The Westminster Press.

## EINSTEIN BIOGRAPHIES

Bernstein, Jeremy 1973. *Einstein*. New York: The Viking Press.

Clark, Ronald W. 1971. *Einstein: the life and times*. New York: Avon Books.

Cuny, Hilaire 1962. *Albert Einstein: the man and his theories*. Greenwich, Conn: Fawcett Publications, Inc.

Frank, Philipp 1947. *Einstein: his life and times*. New York: Alfred A. Knopf.

French, A.P. 1979. *Einstein: a centenary volume*. Cambridge, Mass: Harvard University Press.

Hoffman, Banesh 1972. *Albert Einstein: creator and rebel*. New York: The Viking Press.

Infeld, Leopold 1950. *Albert Einstein*. New York: Charles Scribner's Sons.

Raine, D.J. 1975. *Albert Einstein and relativity*. East Sussex: Wayland Publishers, Ltd.

# Index

118

Edited by Roland S. Phelps